Confronting
the Experts

Confronting the Experts

Brian Martin,
Editor

STATE UNIVERSITY OF NEW YORK PRESS

Published by
State University of New York Press, Albany

For information, address State University of New York
Press, State University Plaza, Albany, N.Y., 12246

Production by E. Moore
Marketing by Dana E. Yanulavich

Library of Congress Cataloging-in-Publication Data

Confronting the experts / Brian Martin, editor.
    p.   cm.
    Includes bibliographical references and index.
    ISBN 0-7914-2913-X (HC : alk. paper). — ISBN 0-7914-2914-8 (PB :
alk. paper)
    1. Business consultants. 2. Consulting engineers.
3. Consultants. 4. Specialists. 5. Expertise. I. Martin, Brian,
1947-   .
    HD69.C6C59 1996
    001—dc20
                       95-19312
                          CIP

10 9 8 7 6 5 4 3 2 1

353353

Contents

1

Introduction: Experts and Establishments

Today's complex society is increasingly dependent on experts—civil engineers, surgeons, taxation lawyers, computer programmers, economists, and many others. These experts are usually defined by their credentials and their solidarity with mainstream professional bodies. Those who oppose them often do not have the same credibility, although they may have the same levels of knowledge and experience.

This book contains first-hand accounts from individuals each of whom has made a challenge to a body of experts. The authors tell about their motivations, their methods, their successes and failures, and the attacks mounted against them. There are some eye-opening stories here, especially in what they reveal about the behavior of establishment experts and the obstacles to open debate. Together, these accounts provide exceptional insight into how to go about challenging the experts.

To introduce this topic, I begin by briefly describing some of my own experiences, before turning to some general considerations. My first major confrontation with experts began in 1976 when I moved to Canberra, the national capital of Australia, and became involved in the campaign against nuclear power and uranium mining. The issue was one of the most prominent of the day: a major environmental inquiry into uranium mining was under way and the government's position was yet to be finalized. As a result, there were numerous media stories. Schools and community groups were eager for speakers. One way I became involved was through the letters to the editor of the city's sole daily newspaper, the *Canberra Times*, which published numerous contributions both for and against nuclear power and uranium mining.

The most prominent and regular pronuclear contributor was Sir Ernest Titterton, Professor of Nuclear Physics at the Australian National University, whose involvement with and advocacy of nuclear technology dated from the 1940s. As a local, high-status authority, Sir Ernest could easily get his articles and letters published. Other prominent pronuclear contributors were Sir Philip Baxter, former head of the Australian Atomic Energy Commission, and Mr. John Grover, a mining engineer.

I composed my letters and articles with care, checking all details with knowledgeable friends.[1] Debate through the letters column was not something for the faint-hearted. I remember the queasy feeling in my stomach the first few times I was directly criticized by later correspondents. How unfair, yet how clever, their arguments sometimes were! There was so much to say in response. Yet, how could I say it all in my next letter, in just a few hundred words, and yet not lose new readers by squabbling over minor details?

Most of the debate was about the role of the civil nuclear power industry in the proliferation of nuclear weapons, the safety of nuclear technology, the economics of uranium mining and the viability of alternatives to nuclear power. The topic of expertise also came up. Sir Ernest asserted that virtually all experts supported nuclear power and that opponents were "a small group of anti-uranium operators who miss no opportunity of spreading their propaganda."[2] Sir Philip presented a more paranoid position, claiming that the antinuclear movement was infiltrated by communists; he was also highly derogatory of individual opponents. John Grover repeatedly made the point that the vast majority of scientists and engineers supported nuclear power, while only a discontented minority opposed it.

The nuclear establishment's argument, that experts know best and that most nuclear experts supported nuclear power, was a challenging one, for it was certainly true that most nuclear experts did support nuclear power. In replying to these arguments, I had one advantage: I was a scientist myself. My recent Ph.D. was in theoretical physics, though not in nuclear physics. But I knew enough science to realize that the nuclear debate was *not* primarily about nuclear expertise. The key issues—environmental hazards, nuclear proliferation, civil liberties in a nuclear society, economics of uranium mining, centralization of political and economic power in a nuclearized world, the impact of uranium mining on Aboriginal communities, and alternatives to nuclear power—involved political, economic, social, cultural, and ethical dimensions.

My response to the "experts-know-best" argument had several strands. First, I pointed out that the so-called experts often had made mistakes in the past. Why should the public trust them now? Second, I argued that expertise in nuclear science and engineering was not central in the nuclear debate. Did knowledge of neutron scattering cross sections really give one a special right to pronounce on energy options? Third, I claimed that the experts had a vested interest in supporting nuclear power, because it was compatible with their careers and world view.

This confrontation with pronuclear experts was illuminating. It was challenging enough for me just to debate the issue through articles and letters in the newspaper. I was very impressed when some of my friends in the antinuclear movement engaged in public debate with Sir Ernest or some other pronuclear speaker. It took real courage to tackle an experienced, self-confident (or, some would say, arrogant), high-prestige scientist in open debate.

There is no doubt that Sir Ernest, Sir Philip, and others did have high prestige in the wider community. Their knighthoods, their eminent positions, and their long influence in government policymaking gave them a big head start in any debate. In the mid 1970s, the idea that Australia's rich uranium deposits should not be mined—when there was plenty of money to be made doing it—was considered radical, if not entirely foolish. Most of us in the antinuclear movement were young and without high formal status. However good our arguments were, we started at a disadvantage in relation to the pronuclear experts.

Things were even more difficult in small country towns. Confronted by a visiting pronuclear expert, the local antinuclear activists were hard pressed to mount an effective response. With an awareness of such situations, I decided to apply my developing social science skills to writing a critique of the views of the leading proponents of nuclear power. An abundance of material led me to focus initially on Sir Ernest and Sir Philip. I tracked down all their articles I could find, using newspaper clipping services, the National Library, abstracting services, and citations. Then I analyzed their views on nuclear power, nuclear weapons, and the nuclear debate. It was no surprise to find that the views of these nuclear experts were closely linked to their professional positions. For example, Sir Ernest and Sir Philip, in the 1960s, admitted a connection between civil nuclear power and proliferation of nuclear weapons because they hoped to keep open the option of Australian nuclear weapons, whereas in the 1970s, they denied this connection since proliferation had become a central argument against nuclear power. My booklet provided a convenient compendium of quotations and critical comment.[3]

My experience in the nuclear debate gave me some understanding of how to go about challenging a body of experts. It also made me aware of how important and how difficult this could be.

The nuclear debate stimulated my interest in the social role of experts, in how experts gain and exercise power, and how they can be challenged. This continuing interest led me to investigate various academic studies of experts, to read many revealing exposés of establishment positions, and to prepare a handbook on methods for challenging experts.[4] But none of these provides much help to those who would like some insight into what it takes to *be* a critic of dominant experts. That is why this book seemed worthwhile.[5] It aims to provide insight into the hazardous business of questioning the dominant experts.

EXPERTS ARE IMPORTANT

It hardly needs mentioning that experts play a crucial role in modern society. If the term "expert" is used in the everyday sense of a person who knows a lot about a subject or can do a task extremely well, then there are experts of all varieties, from bricklayers to brain surgeons and from cooks to computer analysts. Experts in this sense are skilled people.

But there is another sense of "expert" which involves an additional dimension. This occurs when a group of skilled people is able to convince others that they are the exclusive authorities in an area. Bricklayers and cooks have seldom been able to do this: they are rarely quoted in the media concerning policies on housing design or diet. The groups that have succeeded in making their claims to expertise an avenue for considerable power, status, and authority include doctors, lawyers, scientists, engineers, and economists. These occupational groups—commonly called professions—have been able to expand their influence and status beyond what might be expected on the basis of the skills possessed by their individual members. These groups thus can be said to have succeeded in the "political mobilization of expertise," where "political" is used here in the broad sense of involving the exercise of power.[6]

"Political expertise" is a familiar feature of western societies. We are all used to hearing authorities pronounce on various issues. Economists make statements on the economy; doctors make statements about diet. I encountered it in the nuclear debate when Sir Ernest and Sir Philip, on the basis of their position as eminent nuclear scientists, made what they considered to be authoritative statements on energy policy, including fossil fuels and renewable energy sources.

Actually, the preferred role of most experts is behind the scenes, quietly doing their job. Almost all scientists and engineers work for government, industry, or universities. Doctors and lawyers are more likely to have private practices. There are two points that are important here. First, most experts are closely tied to powerful interest groups. Second, these groups are seldom challenged in fundamental ways, and therefore experts do not need to take their case to the public. (There are exceptions to this pattern, however, such as some issues of foreign policy where the experts need to continually present their views and seek to monopolize the discussion.)

Nuclear scientists and engineers worked behind the scenes for several decades—the 1940s until the early 1970s—without having to justify their support for nuclear technology. This was because many governments supported nuclear research, nuclear electric power and, in quite a number of cases, nuclear weapons. When, in the 1970s, a citizens' movement against nuclear power developed, quite a number of these scientists and engineers joined the public debate. They presented themselves as the experts.

This is the usual pattern. Most doctors or civil engineers just get on with the job, most of them working where the pay and conditions are most attractive, committed in their own way to doing a good job. Only occasionally is there some challenge to professional status or conditions: a plan for national health insurance, or the environmental and health damage from a large dam. In such circumstances, a few vocal doctors or engineers are likely to take the lead in defending what they see to be the interests of the profession as a whole.

So here is the general picture: the dominant group of experts in any field is usually closely linked to other power structures, typically government, industry, or professional bodies. The links are cemented through jobs, consultancies, access to power and status, training, and other methods.

Few people would object to such links if the experts were always right. But they aren't. There are many examples where—at least according to later judgments—the dominant experts have backed wrong ideas, dubious or corrupt practices, and illegitimate vested interests. For example, geologists for decades rejected the theory of continental drift. The idea that continents could move was considered eccentric, and those who treated it seriously were treated with suspicion. Yet now continental drift is the accepted theory.

In the early 1930s, in the midst of the economic depression, the standard economic view in industrialized countries was that government expenditure should be reduced. Later economists, following the views of Keynes, saw government intervention as particularly necessary in such times. Military experts provide another example. During the 1960s, U.S. military experts regularly proclaimed that U.S. military involvement in Vietnam could soon be decreased because their communist opponents were nearly defeated. Just as regularly, their forecasts turned out to be completely wrong.

There are certainly plenty of examples showing that individual experts can be wrong.[7] That's only to be expected. After all, anyone can be wrong, even an expert. The important situation is when a whole body of experts is linked to a powerful institution—government, industry, profession, church, and so on—and the expertise is systematically used to serve the institution at the expense of the public interest. When influential experts are wrong in this situation, then it is serious indeed.

This can happen on a regular basis, so long as there is no challenge to the expert claims. An unopposed body of experts has great influence in justifying policies and practices. Enter the critic. When even a single expert disagrees and is able to reach a substantial audience, whether professionals or a wider public, there is no longer unanimity. Instead of an expert monologue, there is now a debate between differing experts. Critics thus have a disproportionate impact on the public perception of an issue. Experts can no longer remain in the background with their positions safe from scrutiny. A few of them, at least, must join the fray to ensure that the critics do not become too influential.

The critics, because they can puncture the appearance of unanimity, often come under attack. They may be slandered, have their publications blocked, or lose their jobs. This may sound extreme, but it is all too common. I started studying the topic of "suppression of intellectual dissent" in the late 1970s. It didn't take long to find that suppression of dissent is a pervasive phenomenon. Indeed, it seems to be a key means by which dissent among experts is discouraged.[8] (The other important means are rewards for conformity—

jobs, promotions, awards—and professional acculturation into a standard picture of the world.)

The contributors to this book are prominent critics of establishment experts. They have taken the courageous and dangerous step of openly and persistently questioning the dominant position. As a result, they have encountered an array of hostile attacks on their credibility and sometimes their careers.

Why are the experiences of these critics worth telling? For one thing, they are simply amazing stories. But, more importantly, society needs more such critics. Without critics, expert establishments have too much power and, as Lord Acton's saying puts it so well, "power tends to corrupt."[9] In order to promote a more open and participatory society, it is crucial that dissident views be heard.

The philosophy behind this book is that society will be better off if more people are able and willing to openly question standard views. This holds true even if critics, by later judgement, turn out to be wrong. What is important is the process of open debate. When debate is inhibited or squashed, the potential for abuse of power is magnified enormously.

It is useful to remember that what we today think of as progress resulted from the overthrow of widely and passionately held beliefs linked to powerful vested interests. The promotion of public hygiene, the abolition of slavery, and the challenge to women's oppression, among others, each took place in the face of powerful forces backed up by esteemed experts.

When I invited individuals to write chapters for this book, I asked them to give a personal account of how they went about confronting establishment experts. Surprisingly, there were few role models I could give them. There are, to be sure, a number of accounts attacking particular bodies of experts, such as Rachel Carson's classic *Silent Spring* and Ralph Nader's classic *Unsafe at Any Speed*.[10] Yet these works give little information about how the critic collected evidence, put it together, and built a persuasive case.[11] There is also a body of academic literature dealing with experts and expertise. But I find it of little use for a practical understanding of what is involved in mounting a critical attack against experts.

When I set about inviting contributors and case studies, I had several criteria. One was the existence of a powerful establishment position with recognized experts or expertise, such as the nuclear industry, orthodox medicine, and mainstream political opinion. Second, I looked for critics who had devoted a major effort to attacking the experts rather than primarily presenting their own particular alternative position. Finally, I looked for cases in which the dominant experts had responded in a way which revealed the nature of the establishment with which they were linked. The contributors and case studies all satisfy these requirements well.

Sharon Beder deals with an engineering establishment that set the parameters for the Sydney sewerage system over many decades. Engineering establishments are incredibly influential in shaping the infrastructure of society: roads, rail, electricity, telephone, water, ports, computer networks, and others. These are not just technical matters: there are questions of power and wealth involved, as well as the direct involvement of corporate and government vested interests. But these political and economic dimensions are usually hidden behind a facade of technical expertise which is seldom considered something for public debate. Beder investigated and exposed the operation of one such engineering establishment, helping to force it, kicking and screaming, into the public eye.

Mark Diesendorf tells about his challenge to the dental and medical experts who support fluoridation. Issues affecting people's health often provoke intense interest and debates, as testified by the prominence of diverse issues concerning cigarette smoking, cholesterol, AIDS, vitamins, and cancer. Experts are involved in these and many other areas, and many of these experts are influenced by powerful interest groups, including pharmaceutical companies, industrial polluters, and the medical and dental professions. Promoters of fluoridation are an especially powerful and well-organized establishment. Diesendorf, one of the world's leading antifluoridation scientists, revealed much about this establishment through his potent challenge to it.

Edward Herman has challenged the scholars, commentators, politicians, and government functionaries who have defined "terrorism" in a way convenient to Western governments. It is a simple fact that most organized killing in the world today is done at the behest of governments, either in wars or by repressive governments against their own citizens. This is forgotten or obscured when terrorism is defined as the action of small antigovernment groups or a few renegade governments. This is one example of how Western governments systematically shape popular perceptions of political reality and are thus able to escape proper scrutiny of their actions. Herman is an eminent scholar and also a committed partisan who has done as much as anyone to expose the double standards of the "terrorism" establishment experts—though this task is enormous, considering the power and ideological sway of national security establishments.

Harold Hillman started off just doing biological research and ended up confronting an enormously powerful biology research establishment. In spite of popular views to the contrary, scientific research is an incredibly conservative enterprise: innovation of particular sorts is welcomed, but challenges to fundamental principles are typically rejected out of hand. The reason is simple: many prestigious and not-so-prestigious scientists have an enormous stake in the prevailing set of ideas and directions. Hillman reveals much about the power of scientific research establishments in his challenge to long-held assumptions about standard methods for biological research.

Michael Mallory and Gordon Moran questioned the standard interpretation of a single art work and thereby came up against the full force of an art history establishment. To some, it might seem that not as much is at stake in the arts as in engineering or government policy, but the same processes apply. Art history is one facet of the more general process of creating and certifying ways of understanding human culture. Various "culture experts" have set themselves up as the authorities in this process, and it is as difficult to challenge orthodoxy here as anywhere else. What is at stake is primarily careers, status, and cultural self-understandings. Mallory and Moran were led into a continuing engagement with an art history establishment which, through its reactions, revealed more about itself than about the art work in question.

Dhirendra Sharma challenged the czars of nuclear power and nuclear weapons in India and, as a result, was targeted for attack. In numerous countries around the world, nuclear technology has been supported by powerful forces in government and industry and opposed by citizen groups. A few experts have had the courage to speak out against nuclear developments and many of them have been attacked for doing so. In India, the task has been especially difficult because of the close personal links between the nuclear establishment and powerful figures in government and industry who had shown their capacity to silence dissent. Another difficulty is the lack of any tradition within India's scientific community of speaking out in the public interest. Sharma paid a serious price for his dissent, but even so he may have been fortunate that the price was not even higher.

I think that each of these critics has a strong case, otherwise I would not have invited their contributions. However, the point of the book as a whole is not to argue that each of these critics is correct and each of the establishments is wrong, but instead to provide insight into the process of confronting an expert establishment, including insight into the operation of the establishment and into successful and unsuccessful methods of mounting a challenge to it.

Reading these accounts, especially the stories of attacks against the critics, makes it tempting to think of expert establishments as unscrupulous conspiracies. Personally, I prefer a different interpretation. Within establishments, the dominant view is so taken for granted that a radically different viewpoint is virtually inconceivable and certainly has no credibility. This means that the critics are easy to dismiss as ignorant or dangerous or both; furthermore, the methods used against them are seen as necessary to protect a worthwhile enterprise. It has long been my view that nearly everyone has the best of intentions, and I believe that the stories told here are compatible with this view. The stories can be interpreted as struggles between groups and individuals each of which believes they are defending or promoting important truths. But some of the contributors may disagree with me on this!

A big challenge faces any expert writing for a general audience: how can the material be made understandable without sacrificing accuracy and rigor? This applies to an even greater extent to critics of experts. (Make no mistake, these critics are experts themselves. They simply disagree with the establishment position.) The views of the critics are much more likely to be unfamiliar to others, and therefore more space is needed for them to explain things, since less can be taken for granted.

As a result, some of these chapters contain difficulties for some readers. Those without scientific training may find parts of Harold Hillman's chapter difficult. Those without familiarity with the visual arts may find parts of Michael Mallory and Gordon Moran's chapter challenging. My advice is to not get stuck on difficult parts. There is plenty of valuable material even for those with no knowledge of the field. Technical detail has been kept to a minimum. For those specialists who want *more* information, plenty of references are cited in each chapter.

There are a number of biases in my selection of contributors. There are numerous critics whose stories would be worth telling and I managed to obtain contributors from a range of fields. Other problems were harder to overcome. A gender balance is difficult to achieve, and would be somewhat artificial, because in many fields most experts, critics or otherwise, are men. For example, virtually every leading figure in the fluoridation debate is a man. Another related bias is my selection of individual critics. Some of the most important challenges to establishment experts come from collective endeavors, most notably within the feminist movement.[12] Yet another bias is my restriction to English-language critics.

To these and other biases I plead guilty. The extenuating circumstance is the importance of the task. I hope that this book will encourage other critics to tell their stories. More importantly, I hope these stories will encourage some readers to become critics themselves and to undertake the challenging and stimulating task of confronting the establishment experts.

NOTES

1. Mark Diesendorf, a contributor to this book on another topic, was especially knowledgeable and helpful.

2. E. W. Titterton, letter, *Canberra Times* (30 March 1979): 2.

3. Brian Martin, *Nuclear Knights*, Canberra: Rupert Public Interest Movement (1980).

4. Brian Martin, *Strip the Experts*, London: Freedom Press (1991).

5. Sharon Beder collaborated in the initial development of the plan for this book.

6. On the analysis of professions as systems of power, see Steven Brint, *In an Age of Experts: The Changing Role of Professionals in Politics and Public Life*, Princeton: Princeton University Press (1994); Randall Collins, *The Credential Society: An Historical Sociology of Education and Stratification*, New York: Academic Press (1979); Charles Derber, William A. Schwartz and Yale Magrass, *Power in the Highest Degree: Professionals and the Rise of a New Mandarin Order*, New York, Oxford University Press (1990); Eliot Freidson, *Professional Dominance: The Social Structure of Medical Care*, New York: Atherton (1970); Terence J. Johnson, *Professions and Power*, London: Macmillan (1972); Magali Sarfatti Larson, *The Rise of Professionalism: A Sociological Analysis*, Berkeley: University of California Press (1977).

7. Christopher Cerf and Victor Navasky, *The Experts Speak: The Definitive Compendium of Authoritative Misinformation*, New York: Pantheon (1984).

8. Brian Martin, C. M. Ann Baker, Clyde Manwell and Cedric Pugh (eds.), *Intellectual Suppression: Australian Case Histories, Analysis and Responses*, Sydney: Angus and Robertson (1986).

9. This insight is confirmed by psychological experiments. See David Kipnis, *The Powerholders*, Chicago: University of Chicago Press (1981, second edition); David Kipnis, *Technology and Power*, New York: Springer-Verlag (1990).

10. Rachel Carson, *Silent Spring*, Boston: Houghton Mifflin (1962); Ralph Nader, *Unsafe at any Speed: The Designed-in Dangers of the American Automobile*, New York: Grossman (1965).

11. Lois Marie Gibbs as told to Murray Levine, *Love Canal, My Story*, Albany, State University of New York Press (1982) is good on this issue but deals much more with establishments than the experts. I thank a referee for mentioning this book and the one by Brint (note 6 above).

12. A prominent example is the Boston Women's Health Book Collective whose book *Our Bodies, Ourselves*, Boston: New England Free Press (1971, and several later editions) constitutes a major challenge to the male medical establishment. The Collective was too busy to contribute a chapter. Their approach is described in, for example, Wendy Coppedge Sanford, "Working together growing together: A brief history of the Boston Women's Health Book Collective," *Heresies*, vol. 2, no. 3 (1979): 83–92.

2

Sewerage Treatment and the Engineering Establishment

SEWERAGE EXPERTS

To be able to relegate the entire job of secondary treatment to a few holes in the end of a submarine pipe and the final disposal of the effluent to a mass of water into which the fluid is jetted, and to accomplish this without material cost of maintenance and none for operation, presents a picture of such great allure as to capture the imagination of the dullest and justify extensive exploration into the ways and means of satisfactory accomplishment.[1]

The sewerage engineers in Sydney, Australia, like many of their colleagues throughout the world, believed that the ocean should be used for sewage treatment. The rhetoric of the Sydney Water Board—that the ocean was "the world's most efficient purification plant"[2]—reflected an attitude that permeated the organization. By the time I began studying the issue in 1985, the use of the ocean for sewage treatment had led to the serious pollution of Sydney's most popular beaches and the heavy organochlorine contamination of fish in nearshore waters.

I had decided that the development of Sydney's sewerage system would be a good case study for a Ph.D.[3] A study of the decision-making processes surrounding the development of Sydney's sewerage system offered an opportunity for me to combine my engineering training and experience with my new interest in the relationship between science, technology, and society. I wanted to find out to what extent technology is shaped by social and political considerations.

I was more interested in studying engineers and engineering than being an engineer but I found my engineering background was not only useful in understanding the engineers I was studying but that it helped in reducing the barriers between us when I began interviewing engineers from the Water Board and the State Pollution Control Commission (the regulatory authority for the State of New South Wales (NSW)). Generally I was accepted as nonthreatening because of my engineering background. Engineers in both the Board and the Commission were quite frank about their views of the public and the role of the engineer, although they were careful about what they said to me about their employers' policies. At no time did any of the people I interviewed at these two organizations admit any misgivings about the ability of the proposed extended outfall scheme to solve the problems of ocean and beach pollution. Nor did they criticize any other Water Board or government policies.

When I began my research the extent of the health and environmental problems caused by sewage in Sydney's coastal waters had been hidden from the public but complaints persisted about the most visible pollution. The Board had begun construction of three deepwater outfalls in 1984 that would extend existing shoreline outfalls two to four kilometers out to sea. The pipes would be laid beneath the sea bottom and the sewage would emerge from a number of diffusers rising from the end of the pipes. The sewage, which contains 42 percent industrial waste, would remain barely treated, with only 10-15 percent of the solids removed. The ocean, the engineers assured everyone, would do the rest.

The Water Board engineers were able to convince many people of this because sewerage engineers were the acknowledged experts when it comes to dealing with sewage. Sewage collection, treatment, and disposal had become part of the professional territory of the engineer in the nineteenth century and despite the divergent fields of knowledge which bear on sewage decisions today, including epidemiology, toxicology, oceanography, marine biology, and many others, engineers have maintained their domination of the area.

Engineers were called in to build and design the first sewerage systems in many European, North American, and colonial cities when the idea of sanitary reform became popular in the mid-nineteenth century. At the time there had been high infant mortality rates and outbreaks of epidemics in many densely populated cities where there was no running water and no reliable and effective means to deal with human wastes. Contrary to popular opinion, the marked increase in life expectancy achieved in these cities during the nineteenth century was not due to the advances of the medical profession, but rather to the engineering works constructed at this time.

In the nineteenth century, whilst engineers designed the pipes, ideas about how to deal with the human wastes, particularly once they had been removed from people's residences, were openly debated by the public and almost anyone could become an expert in the field by studying the issue carefully and writing about it. People from various professions, including doctors and lawyers, wrote books and articles on the subject. This is very different from the situation today when discussion is limited to which engineering solution should be used in a particular situation. Engineers today define what is feasible—what can and can not be done—and which technologies are appropriate. They also attempt to ensure that their preferred solutions are implemented.

The authority of sewerage engineers as a profession, with its own body of specialist knowledge, emerged in the 1870s when two British engineers published books with the term "sanitary engineering" in their titles. This was followed shortly after by an American book.[4] Attempts were made to exclude nonengineers from the field: tradespeople because of their nonscientific knowledge base, physicians because they were unable to execute engineering works, and public health officials and municipal bureaucrats because they did not have sufficient breadth and depth of training. Sanitary engineers were to be civil engineers with additional knowledge of physical and natural sciences.[5]

The aptitude of engineers, however, particularly with respect to sewage treatment, was not immediately apparent, even in the nineteenth century. Sewage treatment involved biological and chemical processes that scientists and others felt they had a claim to. For example, towards the end of the nineteenth century some scientists, biologists in particular, threatened to take control of sewage farming as the biological mechanisms of sewage farming became better understood.[6]

Despite the enormous popular appeal of sewage farming in the nineteenth century and to the present day (because it makes agricultural use of the nutrients and water in the sewage) engineers were not inclined to favor it as a method of treatment because it was unpredictable, less controllable, and less closely aligned to their traditional skills than more artificial methods of treatment. The "naturalness" of a sewage farm, which appealed to some sections of the public, was not a desirable attribute to engineers who sought to harness and control nature with their technologies and thereby claim expertise in sewage treatment.

The triumph of engineers in taking control of sewage treatment marked an end to sewage farming as a feasible treatment option in most Western countries. As a group, sewerage engineers (also called public health engineers and sanitary engineers) preferred certain technologies and methods and virtually ignored others. They favored water-carriage methods to transport the sewage to the nearest waterway for disposal despite considerable public opposition in some cases. Treatment methods were developed to ensure that the sewage, when discharged into a river, would not use up too much oxygen and choke the river or create a nuisance because of a build-up of rotting matter. A large variety of treatment processes were soon reduced down to a manageable few that were arranged into primary, secondary, and tertiary stages.

Where engineers were able to use the sea for disposal they avoided treating the sewage altogether because the sewage would be diluted and decomposed in the sea. Engineers always sought the cheapest solutions and preferred not to install treatment if the ocean would do it for them. Their defense of this practice was initially quite flimsy and unconvincing but over the years it has become more sophisticated with the addition of complex oceanographic studies which are designed to show that sewage will be diluted, dispersed, carried away by currents, remain submerged beneath the ocean surface, oxidized and treated by the ocean, and generally rendered harmless.

ENGINEERING PROOFS

In 1985, I became interested in how Water Board engineers had managed to consistently claim for decades that sewage was being carried away by a southerly current and diluted and decomposed far away from the beaches when it seemed obvious to anyone who watched the movement of the sewage fields that the sewage often came onto the beaches. And it must have been obvious to those who went swimming on such occasions.

My historical research showed me that even before the first ocean outfall had been built in 1889 people who lived near the sea had seen garbage and offal disposed of at sea come back onto the beaches. The letters to the newspapers at the time showed that many people did not believe the engineers' claims that the proposed outfall near Bondi Beach would not cause any pollution of the beach.

The engineering textbooks and expert writings of the late nineteenth century indicate that engineers were well aware of the fact that sewage would rise to the surface of the ocean (because it had a higher temperature and lower specific gravity than sea water) and flow in the direction of the wind, which could be onshore.[7] The textbooks recommended that engineers study the currents and tides with the use of floats. But I found it puzzling that they instructed that the floats be kept submerged so that they would not be affected by the winds.[8] If the sewage traveled in the direction of the wind, why did they not want the floats to travel in the direction of the wind?

In Sydney, engineers used such float studies to argue that the sewage would be carried away by the southerly current even though there was a predominant onshore wind in summer. After the first ocean outfalls had been built evidence that the engineering predictions had been wrong inevitably emerged. As the sewage comes to the surface of the sea it forms a field with sharply defined edges which can be differentiated from the sea water by its discolouration. The fields can be observed to travel in one direction or another from adjacent headlands and if onshore winds are blowing it is easy to trace their course onto nearby beaches. Other signs of pollution are also readily visible. Floating solid material in the water and grease balls on the sand are two obvious examples. Smell and greasy feel are other good indications of the presence of sewage.

Despite these obvious indicators, for almost one hundred years the Water Board engineers persistently denied that the pollution resulted from those outfalls. They explained that the sewage could not have come from the outfalls because of the southerly current that would have carried it away. Their theoretical predictions were given more weight than the real evidence that contradicted them. From then until the present day, the obvious pollution was blamed on passing ships, algae, beachgoers, and stormwater drains. It is difficult to understand such denials except in terms of deception of the beachgoing public. And such deception relied on the authority of expertise to gain the support of the wider public.

In the summer after I began my research, the Water Board spent half a million dollars telling the public how its deepwater outfalls were going to clean up the beaches. The deepwater outfalls would, they said, end sewage pollution of the beaches (which the Board now belatedly admitted came from the outfalls). This message appeared in television and magazine advertisements that were always visually splendid. Pristine beaches and beautiful people evoked a promised future of unpolluted beaches.

As in the past, it was claimed that the sewage would be carried off by the southerly current and treated in the ocean through dilution, oxidation, and biodegradation.[9] It all sounded rather similar to the historical material I had been reading. But there was a new element this time. The Board was predicting that the deepwater outfalls would keep the sewage field submerged. If the sewage was submerged beneath the surface of the ocean, it would not be blown in the direction of the wind.

The Board's claims were based on the existence of a thermocline, or difference in densities, in the coastal waters off Sydney. The idea was that sun-warmed waters on the surface would be less dense than the cooler, deeper waters and therefore would not mix with them. Sewage released into those deeper, denser waters would be trapped beneath the surface of the ocean, under the layer of warmer water and carried southwards by the ocean current. This theory had originated in the United States when it had been discovered that the Los Angeles deepwater outfall seemed to work this way most of the time.

However, it was apparent that conditions off the coast of California were different from the conditions in the coastal waters off Sydney. For one thing, the current coming from the north of California came from the cold northern regions and provided more difference in temperature and therefore density from the sun-warmed layer on top. In Australia the northern current came down from tropical waters and so was much warmer to start with. Would the difference be enough to keep the field submerged off Sydney? And if it was, would the current really carry it all away?

One person who claimed it wouldn't was Tom Mullins, a marine chemist at the University of Technology, Sydney. He said that there was no single unified south-going current off Sydney but rather a series of eddies and other irregularities. A wording change in a Water Board's public relations brochure also made me suspicious. An early brochure stated that:

> the effluent/seawater mixture moves away from the initial dilution zone under the influence of ocean currents. In Sydney, these currents are not normally directed onshore during the summer months.

A reprint of the same brochure was changed to:

the effluent/seawater mixture moves away from the initial dilution zone under the influence of strong offshore ocean currents during the summer months.

I examined the oceanographic studies undertaken by the Board's consultants, Caldwell Connell. In two studies, one in 1976 and one in 1980, Caldwell Connell had measured currents at various depths where the outfalls would be discharging. These studies showed that even the deepwater currents were going towards the shore for 30-50 percent of the time during the summer. Yet they ignored their own evidence and concluded that in the long term the sewage discharges would be carried southward.[10] I questioned the Board's engineers on this point and was told that although the currents were going towards the shore, they turned when they got close to the shore and headed southwards. This assumption was based on theory but had not been tested empirically.

Even if you accepted this, there were significant differences between the claims in the engineering reports and those being made by the advertisements. For example, while the advertisements said sewage pollution would be eliminated, the reports predicted that the sewage fields would still come onto the beaches when the field surfaced and there was an onshore wind. This would happen, the reports stated, for a small amount of the time in summer and 40 percent of the time in winter, when many people still go to the beach and surf.

My suspicions were further aroused when I discovered a retired Commission scientist, Robert Brain, who had studied the Water Board's models in detail and who argued that they were wrong. Brain had given an honest appraisal of the Water Board's predictions when asked by his superiors at the Commission but now realized that was not what they wanted. He claimed he had subsequently been victimized, moved sideways, and his career ruined as a result. He told me that at one stage, whilst he was away on holidays, his personal files were gone through and some material removed. Commission engineers, whom I interviewed, tried to discredit Brain but I discovered during the course of my research that Brain had actually been highly thought of by senior Commission engineers before he questioned the Water Board's predictions.

This discovery happened one day when I was researching in the Commission's offices, and had asked to look at some files from a few years before. They were brought up from the bowels of the building by a junior officer and placed on a desk where I set to work, reading through them and taking handwritten notes. In the files I found a 1980 memo by the Principal Engineer for Water, Wastes and Chemicals saying that he believed that there were only two Commission officers with the necessary expertise to undertake the assessment of the Board's models and calculations and Brain was one of them. Another memo from the Principal Engineer stated that he could not find anything wrong with Brain's criticisms and that the Commission should not ratify the Board's proposals until the issues raised had been resolved. "Otherwise in the event of a public inquiry, the Commission might justifiably be subjected to serious criticism."

I was feverishly writing all this down when a senior Commission officer who had been observing me, came up to me and asked me to stop as I had been given the files by mistake. He said he needed to consult the Commission lawyer about whether I could look at these particular files. He tried to explain that Brain had been discredited and that the material I was looking at was not relevant. He took the files away and asked me for my notes. When I refused he wasn't sure what to do and let me keep them. After that I was only allowed to see parts of the files that had been given prior approval for my perusal by the Water Board.

VESTED INTERESTS

This sort of behavior only encouraged me to delve deeper. Why was the Commission protecting the Board? They were the regulators and were supposed to be concerned about public health and environmental protection. Why were they so committed to the deepwater outfalls? Was it because both organizations were dominated by engineers? Were they subject to the same pressures from their political masters? Some more historical research allowed me to see another part of the puzzle.

When the Commission had been formed in 1972 it had been charged with implementing the Clean Waters Act and cleaning up Sydney's waterways which were severely degraded with industrial waste. The Commission had achieved this feat by requiring firms that were discharging their wastes into the rivers and creeks to divert their wastes into the sewers. The Water Board obliged the Commission by allowing those firms to do this and in this way the industrial waste was removed from the rivers to the ocean. The

Commission was somewhat beholden to the Board for this and could hardly turn around and penalize the Board for the huge quantities of toxic materials that were now pouring into the ocean from the Board's outfalls nor for the resulting marine pollution. Instead it actively helped the Board to keep knowledge of the resulting fish contamination from being made public.

I found that the Commission placed no formal restrictions on what toxic material the Board could put into the sea. The guidelines for toxic materials were expressed in concentrations in the environment rather than total amounts. When I multiplied the concentrations by the actual flows and claimed dilution factors I found that under the guidelines the Board could have discharged huge quantities of heavy metals and organochlorines, in some cases more than the total amounts produced in New South Wales. Even so some substances, particularly organochlorines, were approaching those limits. However, the deepwater outfalls would ensure further dilution of the sewage and meant that the amounts of toxic waste that could be discharged under the guidelines, when the deepwater outfalls were built, would increase dramatically.

When the choice was made between upgrading sewage treatment onshore or building deepwater outfalls that would delegate this task to the ocean, the Board had chosen deepwater outfalls and the Commission, after seeking advice from one of their consultants, a retired engineering professor, agreed. Upgrading the treatment to secondary treatment would not only have been more expensive but would have required restrictions on industrial waste being allowed into the sewers because secondary treatment utilizes naturally occurring microorganisms that are sensitive to toxic waste. The deepwater outfalls enabled industry to keep using the sewers as a cheap toxic waste disposal service.

The problem with using the ocean to treat the sewage is that most people don't like swimming in a de facto sewage treatment plant because they think it might be unhealthy. The Commission told the public that coastal waters could be presumed to be bacteriologically safe for swimming if aesthetic criteria were met; in other words, no undisintegrated fecal matter or other materials "clearly of sewage origin" should be allowed into bathing areas and also no "noticeable" turbidity or discoloration of bathing water attributable to sewage and no "perceptible smell." After some delving I discovered that this view was based on a 1959 study undertaken in the United Kingdom. I found that this study was still referred to in Britain, Australia, and New Zealand as the classic paper on the subject yet it didn't take much research to uncover a continuing debate among international experts on the extent to which sewage polluted water posed a health hazard. There was also plenty of more recent research and developments in the field of virology reaching conclusions contrary to the 1959 study. For example, I discovered that epidemiological studies in the United States since that time demonstrated significant risks of bathing-associated disease in recreational waters that are mildly contaminated with sewage. In 1980, a U.S. EPA spokesman claimed that

> surveys of 30,000 bathers and non-bathers contacted on beaches in New York and Boston revealed statistically significant increases in cases of vomiting, diarrhoea, nausea, fever and stomach aches among swimmers who had bathed in polluted waters. . . . The results show a strong link between bacteria counts in the water at the time of bathing and subsequent health of the swimmers.[11]

I also tracked down a paper given at an International Conference on Water Quality and Management for Recreation and Tourism in 1988 which summarized data collected by the New South Wales Health Department between October 1983 and April 1987. Salmonella was detected in 183 out of 1058 (17%) samples tested at Sydney's eastern suburbs swimming spots and beaches. Moreover, the Health Department's monitoring of bacteria levels at beaches found that the same beaches were unsatisfactory for swimming for between 29% and 83% of the time, depending on the beach and whether it had rained in the previous twenty-four hours (when the sewers overflowed and the treatment plants were bypassed).

However, during this time the Water Board's Annual Reports showed that the beaches were meeting standards most of the time. How could this be? The standards the Board was referring to were standards set by the Commission which were different in significant ways to the Health Department criteria. I found that the Health Department classified beaches satisfactory for bathing or unsatisfactory on particular days while the Commission standards used a statistical measure that allowed days of heavy pollution to be covered up. I was able to use raw sampling results from the previous summer to show that whilst the Commission standards were being met some beaches were in fact unsatisfactory for swimming according to the Health Department for half the time.

I could find no record of the Health Department telling the public of its contrary findings or undertaking any sort of study to find out what the implications of their sampling were in terms of human health. As far as I knew, and this has been confirmed since, no epidemiological study had been carried out of swimmers in Australian waters. Without such a study, the Water Board and the Commission were able to continue claiming that beaches which met Commission standards were not a health hazard.

Although many beachgoers knew the beaches were polluted and that they occasionally got sick from swimming, the government experts were seldom challenged by outside experts, either engineers, doctors, or scientists. I found that formal complaints had been regularly made by the beachside councils behind the scenes but the councils were reluctant to take any public stance that might advertise the fact that their beaches were polluted and thereby turn away potential residents or visitors and beneficial business in the area.

LACK OF EVIDENCE

If you're going to use the ocean for sewage treatment, it seemed to me that it was vital to consider the fate of viruses and toxic materials that enter the ocean in this way. Yet over more than a decade while the Board's consultants, Caldwell Connell, undertook their detailed studies of Sydney's oceans, these areas were almost completely neglected. Their million-dollar feasibility study[12] was heralded by the Board as the most comprehensive study of its kind ever undertaken in Australia. They studied the biological characteristics of marine life in some detail, examined the composition of the water and its concentrations of oxygen and nutrients, and they mapped out the topography and geology of the coastal region. But they did not study viruses, pathogenic bacteria, nor the toxic content of marine life in the area.

Viruses, Caldwell Connell said, were difficult and costly to test for and testing could not be carried out without specialist assistance.[13] So why didn't they get that specialist assistance as they had in other areas? Sewerage engineers recognize the limits of their knowledge and increasingly draw on the expertise of environmental scientists and others, by hiring them, using them as sub-consultants, or drawing on their literature. But this use of other experts is often subordinated to their own ends. I found the engineer-dominated government authorities and the engineering firms they worked with were highly selective in their usage of other experts, often drawing on them merely to justify their proposals and cover their failings, or not using them at all as in this case.

Having admitted their lack of expertise in the area of viruses, Caldwell Connell assumed that viruses would not live long in the ocean and their numbers would "diminish rapidly through treatment, dilution and natural die-off."[14] Yet the textbooks said that the treatment Sydney sewage received would not reduce the numbers of viruses and I uncovered several studies that showed that viruses could live for months in sea water, whereas the fecal coliform[15] that Caldwell Connell did study die off in a matter of hours. Caldwell Connell admitted that there was very little evidence that related "faecal coliform concentration to the incidence of water borne disease" but studied their die-off rates "as a matter of convenience."[16] I found this extraordinary. How did they get away with it?

Their study of the fate of toxic material was similarly lacking. I know that while organic matter does eventually decompose in ocean water, heavy metals and organochlorines tend to persist in the environment, accumulate in seabed sediments, and bioaccumulate in the food chain. Yet this possibility was not properly investigated by Caldwell Connell who stated in their feasibility study that "a detailed investigation of levels of pesticides and heavy metals in the marine environment is beyond the scope of this study."

In the environmental impact statements,[17] which were also prepared by Caldwell Connell, the possibility of bioaccumulation of toxic substances was dismissed as unlikely since no serious accumulation of these toxic materials had been observed in sediments near the existing outfalls. But I found they had hardly even looked for sediments. They had only taken samples in three places for analysis of toxic contamination and these were some distance away from the existing outfalls. In a confidential report that I uncovered, the Commission noted that "The statistical significance of single samples and the validity of a sampling technique which does not segregate undisturbed surface material must be brought into question."

Caldwell Connell assigned no importance to the fact that elevated levels of heavy metals and DDT were found in the sample taken nearest to the largest outfall at Malabar and argued that this material "appeared to be deposited only during periods of low current velocities and was dispersed under the normal current regime." These meager observations were sufficient justification for Caldwell Connell to assume that toxic material did not accumulate, despite the obvious evidence that it had.

By studying the responses of government departments to the environmental impact statements I found that they were less optimistic than the Board and its consultants. A major concern of the Department of Mineral Resources was the potential accumulation of deposits of solid particles which might in turn lead to a concentration of heavy metals and toxic chemicals. They were sceptical of the claims that ocean current velocities/settling times/particle sizes were such that wide dispersion of solid particles would occur. "It is difficult to understand that these particles do not go somewhere specific where they accumulate."

The Australian Museum, which had conducted ecological surveys of nearshore waters for the Water Board, claimed that particles from the diffusers which fell into the mud/clay range would be likely to be deposited in a relatively stable region of mud and that heavy metals and other industrial wastes which might behave like mud or clay sized particles were likely to also be deposited in this stable zone of muddy sediment. Such materials could then be assimilated by benthic organisms and enter the tissue of fish passing through the area. "Such a situation could be harmful since the professional fishing grounds of Sydney are located in this region."

I knew of at least one survey that had been done of fish contamination while Caldwell Connell had been conducting their massive feasibility study and this showed that heavy metals were accumulating in the fish. I uncovered a Caldwell Connell internal report of a meeting to discuss the survey, published well before the completion of the feasibility study, which stated that "It was agreed that, while the data only represented analyses of individual specimens, levels of heavy metals and pesticides detected in this small number of samples were such as to suggest that a potential public health threat or environmental hazard might exist within the study area. . . ."[18] Yet there was no mention of this in the feasibility study and no further surveys undertaken by Caldwell Connell. The only public report of the fish survey that I could find was in the 1979 environmental impact statement which stated: "Whilst the statistical significance of the 1973 survey is not able to be clearly established the results are encouraging in that they indicate that no serious environmental problem existed even prior to the full implementation of source control of restricted substances. . . ."[19] A very different public interpretation!

On the whole I found fish contamination reports the hardest to obtain. Some had been done by the Fisheries Research Institute (part of the Department of Agriculture) and never published. These were not mentioned in the environmental impact statements despite their direct bearing on the environmental impact of the proposed deepwater outfalls. If toxic material was accumulating in fish life, as the few reports I managed to get hold of indicated, then the deepwater outfalls would be putting the same toxic material further out to sea, closer to the commercial fishing grounds. One study of pesticides in fish caught near the outfalls[20] seemed impossible to track down, even with the help of a member of parliament. It had been carried out in 1979 but was not published until 1989 after I told the newspapers of its existence. Needless to say, it revealed contamination of fish above Australian standards.

MY INVOLVEMENT

As I neared the completion of my research it was obvious to me that there was a major sewage pollution problem in Sydney waters that had largely been covered up by the experts and the organizations they worked for. It also seemed that the deepwater outfalls, far from solving the problem, were likely to cover it up even more by removing some of the visible evidence of the pollution. It seemed imperative that I go public with my findings. While I was undertaking my research I had refrained from making public statements about the sewage pollution or the outfalls. Instead I kept the environmental group Stop The Ocean Pollution (STOP) informed so that they could campaign against the outfalls more effectively.[21] (STOP was a small group of beachgoers, surfers, and environmentalists.)

It became evident that the Water Board suspected this connection when, at one interview, Water Board public relations people confronted me about my environmental affiliations, producing notes of a talk I had given at a seminar at the University of Wollongong which had been attended by about twenty people. They said that I had apparently been influenced by members of STOP. They told me that I should not listen to them because they didn't know what they were talking about. They proceeded to put me right about what a good job the Board was doing.

It was partly because the Board's public relations people were so successful at undermining STOP's credentials that journalists were reluctant to report their statements regarding the scientific basis of the extended outfalls. They repeatedly asked STOP members if there were any 'experts' that they could refer them to.

My decision to speak out was not a difficult one. In many respects, although I was trained as an engineer, I have been far freer than most to challenge my fellow engineers since, as an academic and writer, my career prospects are not dependent on endearing myself to the engineering profession or gaining employment in an engineering firm or government department. I could understand that engineers and scientists employed by the Board or the Commission, who might have felt uncomfortable with what was going on, could not speak out because they were concerned about their jobs. One or two seemed very nervous just talking to me.

There were rumours about engineers who were critical of the proposed outfalls but they remained well hidden. Back in the 1970s, when the idea of the deepwater outfalls was fairly new, the *Daily Telegraph* had reported that "private and government civil engineers" had criticized it arguing that it would do little to solve the pollution problem.[22] Such critics had not been willing to put their names to their criticisms, however. Most sewerage engineers in Australia are employed by government departments or instrumentalities and those who aren't are consultants dependent on those same government departments for work, or academics dependent on them for research funding. So critics face the possibility of severely limiting their career prospects. Those engineers who are not employed as sewerage engineers still face disapproval and censure from the engineering profession. It is an unwritten part of the engineering ethos not to criticize works designed by other engineers, because this may reflect badly on the profession.

John Tozer, a structural engineering consultant, found this when he criticised a proposal to build an outfall near his home at Look-At-Me-Now headland in northern New South Wales. In 1990, he was found guilty of breaching the engineer's code of ethics because of his public criticisms of the local council engineers who supported the scheme. He was subsequently eased out of the Association of Consulting Engineers, Australia (ACEA). Recently, Tozer was publicly admonished by the Institution of Engineers, Australia (IEAust) for failing to uphold the honor and dignity of the engineering profession because he used "intemperate language" in a private letter he wrote to the Premier that criticized the outfall. The letter was on his business letterhead and identified him as an engineer.

After I began to be quoted in the newspapers in 1989, I too was accused of breaching the engineering code of ethics. I was phoned one evening by a senior member of the Institution of Engineers and accused of not upholding the dignity and honor of the profession and speaking outside my area of competence (despite the fact I had just done a doctorate on the subject). The caller threatened to make a formal complaint against me.

The Institution of Engineers also sought to publicly support the Water Board engineers. Its president issued a press release that read, in part:

> I deplore the denigration of Australian engineering endeavours which seems to occur too frequently these days. Innovative projects of this type should be recognised and supported by the community.
>
> Australian engineering ability and performance is recognised throughout the world as being of the very highest calibre, with the Water Board in Sydney having its share of distinguished engineers. It is important to Australia's competitive performance that, where deserved, Australian engineering excellence is supported by our mass media. I believe the Ocean Outfalls project deserves this support.

It was later revealed that this man's consulting firm had been retained by the Water Board as management consultants on the deepwater outfall project.[23]

Nevertheless, after this initial reaction, the Institution of Engineers itself has sought to hear and incorporate my views. I was invited to join the Institution's Environmental Engineering Branch committee the following year, at the suggestion of the same President. (I have been a member ever since and became chairperson in 1992.)

The Institution's magazine, *Engineers Australia*, did a feature story on the outfalls controversy in February 1989, which covered my views fairly and promoted some discussion in the letters section. One letter said "it has continually amazed me that the debate is being carried out by laymen with nary a word from the professionals. Of all the people who should be able to provide information to the public, civil engineers are the best placed yet are noticeably silent."[24] (It had obviously escaped his attention that I was a civil engineer.)

When my book *Toxic Fish and Sewer Surfing* was published later in the year,[25] the editor of *Engineers Australia* gave it a favorable review. I received a much less favorable review in the magazine of the Australian and New Zealand Association for the Advancement of Science (ANZAAS), *Search*. It was written by a government scientist who was an inventor of a sewage treatment process I had criticized. He called the book a "polemic against engineers in general and the Sydney Water Board in particular" and stated:

As one who has watched the events from a safe distance, it is clear to me that the debate has been maintained at an emotional level, with a propensity for exaggeration and limited vision being displayed by both sides. While I found the book both entertaining and informative, the more I read the more uncomfortable I became, as the tone became shriller and the close personal involvement of the author with the issue more obvious.[26]

I was subsequently invited to speak at an ANZAAS Seminar on "Sydney's Strangled Sewerage System" and later to speak at the Institution of Engineer's Annual conference. I was well received at both although subject to some angry questions, particularly from Water Board employees and sympathizers.

CONTROLLING THE INTERPRETATION OF INFORMATION

During the course of my research at the State Pollution Control Commission, I had come across some figures for levels of organochlorines in fish caught near Sydney's main outfall at Malabar. They were in the business papers for a meeting of the Clean Waters Advisory Committee which was a committee of representatives of various government departments and government appointees representing selected interest groups which advised the Commission.

I copied the figures down without knowing their meaning since there were no standards included with them but when I later compared them to the National Health and Medical Research Council (NH&MRC) maximum residue limits I was astounded at how high they were. Two out of three species of fish which were tested were over the NH&MRC limits and one species, the red morwong, was over the limits for benzene hexachloride (BHC) by 122 times on average (eight samples of each species were tested) and over the limits for heptachlor epoxide by an average of over fifty times.

When Alan Tate, from the *Sydney Morning Herald*—Sydney's quality daily newspaper—interviewed me a couple of months later for a story he was writing on sewage pollution, I pulled out the figures to prove to him that there was already a pollution problem caused by industrial waste being discharged through the sewage outfalls. Tate was keen to publish the figures but could not rely on my notes alone. He needed to authenticate them. He made his own inquiries and finally found one person in the Commission who said he could confirm the figures. The person then changed his mind because he was concerned about his job. Tate rang him several days in a row without success until the day before the figures were due to be published. Tate suggested to the person that all he had to do was cough if the figures were correct. He coughed.

Tate then rang the Water Board to tell them he had the results of the study and was warned by a senior Water Board officer that the *Herald* should not publish the results because they were open to question. The officer said that the Board thought the BHC, found in such high levels in the red morwong, might really have been lindane (a specific form of BHC) and therefore only slightly over the limits. (The maximum residue limits for lindane are much higher than for other forms of BHC). Tate then rang the Australian Analytical Laboratories, which had performed the analysis, and was told that there was no doubt that the substance in question was non-lindane BHC.

The results of the study, which had been done in 1987, were published for the first time the next day, 7 January 1989.[27] Not only had the Commission kept the results of the study secret for more than a year, but several other government departments, through their representatives on the Clean Waters Advisory Committee, had known of the findings. Yet there had been no leaks. It later emerged that the Minister for the Environment had instructed the Commission not to reveal results of the study in its Annual Report.[28]

It also emerged that Water Board officers had met with Commission officers in May 1988 and a memo of the meeting stated that "spearfishermen consuming red morwong caught at Malabar could be at some health risk."[29] Yet the results were not even given to the Australian Underwater Federation when it wrote to the Commission in September 1988 asking for results of the study. Their letter said that their members, including spearfishermen, had noticed that red morwong caught near the outfalls had mushy, tainted flesh and they were concerned about whether they were safe to eat.[30]

The Board's planning manager later defended their decision not to inform the public of the results:

> The criticism that by withholding the study results the board was potentially putting public health at risk had to be weighed up against the risk of causing unwarranted public concern and panic.[31]

The significance of the study was that the Water Board engineers had been claiming for several years that toxic industrial waste did not accumulate in the marine environment near Sydney and therefore the extension of the outfalls would not cause a pollution problem in deeper water, closer to the commercial fishing areas. Several previous studies that had also shown accumulation of organochlorines (particularly dieldrin and DDT) and heavy metals (particularly mercury and cadmium) in fish caught near the outfalls had also been suppressed.[32]

With full knowledge of all these results the Board went ahead and published an advertisement for the extended outfalls which referred to the ocean as "the world's most efficient purification plant" and stated:

This is also the world's largest and most natural treatment plant, and it has some of the most experienced employees as well. Hundreds of species of fish and other marine organisms exist here to do little more than thrive on breaking down the pre-treated effluent discharged into the ocean off Sydney.[33]

Even after the leak to the *Sydney Morning Herald*, the Water Board, the Department of Agriculture (responsible for fisheries), and the Health Department continued to down play the significance of the study. Water Board officers claimed, "The results obtained from this study were from a very small sample number and were not compared to any sample with a known concentration. It is not unusual for studies of this nature to have high errors associated with them due to natural variations within the sample population."[34] They claimed that the large amounts of heptachlor epoxide found in the study were really a sulphur compound.[35]

A second study, that had been carried out in 1988, had sampled red morwong at varying distances from the three major outfalls and included a comparison of four different laboratories so as to meet criticisms of the Australian Analytical Laboratories, which had done the analyses for the first study. The study concluded that only that laboratory and one other accurately detected a wide range of organochlorines. It showed different organochlorines accumulating in the fish above the NH&MRC limits, particularly chlordane and hexachlorobenzene (HCB).

This interlaboratory study raised even more disputes. The Minister for Agriculture wrote to the Minister for the Environment, after both studies had been reluctantly released in March 1989, to express his concern about the continuing publicity being given to the contamination of fish. He argued that very small errors in technique or measurement could seriously flaw the results when measuring minute amounts of chemicals in fish. He argued that reports of both studies were potentially erroneous because they had not been refereed "in the standard scientific manner,"

> I would appreciate it if you would ensure that media reporters are fully aware that these reports do not have the scientific standing that is being attributed to them . . . we should take all possible action to prevent the continuation of the unsubstantiated reporting which is doing so much needless damage to one of our State's most important industries.[36]

An independent referee's report, subsequently procured, generally approved of the studies saying that the "basic nature of the problem has been adequately identified and evaluated."[37] Another review was made by the Director of the Southern California Coastal Water Research Project. He had no major criticisms of the studies. He agreed that both showed that red morwong were contaminated near the outfalls. He suggested, as a public relations strategy (and as a way of shaping perceptions of the meaning of the results):

> After evaluating the best world-wide evidence for health risk from the various organochlorines, you might want to release to the press a comparative table to put the risks in line with others commonly accepted by the public.[38]

In a different report he advised the government:

> The recent events in Sydney indicate a route of communication to the public from the scientists should be developed. This may reduce the "scare" from the press and shield the fishing industry from impacts produced by false or inaccurate media reporting.[39]

In the past the Board could be fairly confident about getting its press releases published and its version of events reported. The Board's public relations department had a comfortable relationship with the media, putting out the occasional brush fire with their version of the facts, and that version was almost never challenged by journalists. 'Serious' papers like the *Sydney Morning Herald* could be relied upon not to report unsubstantiated claims and to give preference to 'expert' opinions from government officials.

But in 1989, things had changed. The government and its advisers were well aware that the *Sydney Morning Herald* journalists and some television journalists were coming to me for interpretation of data, reports, and anything else that they found or that the government released that had to do with sewage pollution. Alan Tate from the *Sydney Morning Herald* had originally been referred to me by the local Friends of the Earth office. He found that I was a reliable source of information and continued to come to me. Other journalists soon followed and I became one of the people that journalists sought to provide an environmental point of view. STOP members were also now credible sources of information for the media and part way through 1989, STOP purchased a fax machine through which they could put out media releases. These releases were fairly successful at gaining news coverage now that STOP had become known to the media.

The government sought to avoid alternative interpretations by imposing their own at the time of release of reports, particularly those likely to be damaging. Shortly after the results of first bioaccumulation study were published in the *Sydney Morning Herald*, a reporter asked a senior Board scientist whether figures given by the Board for concentrations of toxic substances in discharged effluent in one of their recent reports included the portion of these toxic substances in sludge which was also discharged into the ocean. He was told by this scientist that of course they did, and "You don't think I would let them be published if the sludge was not included do you?" I attempted to prove to the reporter that the Board's spokesman was lying and he made further inquiries at the Board. He was told by a puzzled public relations officer that the Board's engineers were rushing round and that the Board was in a state of chaos. The next day the same scientist admitted that the sludge figures had not actually been included in the report and that "an honest mistake" had been made in telling the reporter otherwise.

That same week Ian Wallis of Caldwell Connell came up to Sydney and the Board held a press conference. The *Herald*'s reporter, Alan Tate, claimed that virtually everything Wallis said during their meeting "was useless as far as reporting the issues at hand" and Paul Bailey, the paper's environment writer, said that if they had reported the meeting they would have reported Wallis's admission that further treatment would have to be investigated eventually for the ocean outfalls. Yet many Water Board people were incensed that Wallis, the expert, did not receive any coverage in the *Sydney Morning Herald* and took this as a further sign that the *Herald* was biased and was conducting some sort of vendetta against them.

The relationship between the Board and the *Herald*'s key sewage pollution investigators continued to deteriorate. In a subsequent screaming match between Tate and the Water Board's public relations manager, the public relations man alleged that Bailey had admitted in the meeting with Wallis that he had no understanding of how the extended ocean outfalls worked and hadn't realized the complexity involved. Tate denied this but the same story was reported in the Board's internal magazine, *Aquarian*, except that this time it indicated that it was Tate who admitted he had no understanding of the project. Tate was incensed but no doubt it did wonders for the morale of Water Board employees who felt besieged by the *Herald*.

In the meantime, the Board had given the *Herald* some figures for concentrations of toxic substances in sludge but in a form that was difficult to interpret. I studied the figures, comparing them to other information I had, and decided they were not credible, but I was unable to prove them to be false. A few days later, the Commission admitted that it didn't require the Board to monitor the sludge for toxic material and the Board's monitoring manager was reported as saying that the Board did not know much about the concentration of toxic material in sludge discharged through its outfalls. "We have started looking at this in the past few weeks" he said.[40]

At the beginning of March, the Board placed a full page advertisement in the *Herald* which claimed that 70 percent of the volume of industrial discharge which could contain damaging wastes had been controlled by their Trade Waste policy and that this meant that "this waste may no longer be discharged to the sewers or drains." Such a statement was patently untrue. Neither the 70 percent of industrial discharge nor the damaging wastes were prevented from entry into the sewers under the Trade Waste Policy. I pointed this out to journalists and the Board was forced to admit, shortly afterwards, that the advertisement was "certainly ambiguous" and "should be clarified."

The *Sydney Morning Herald* had begun labelling its articles on the issue "Sydney's Watergate" and the Minister tried to reassure a press conference that there would be no more cover-ups or lies. Indeed, he said: "We are determined the Water Board will tell the truth, the whole truth and nothing but the truth and if that requires, as Richard Nixon once described it, minor correctional statements as we go along, then they will be taken."[41]

The Board also tried a number of ways to stem the bad publicity. Apart from direct threats of legal action it was rumored that Water Board executives were lobbying senior executives of the *Sydney Morning Herald* to get the series of damaging articles stopped. Early in the piece the Board's managing director accused the *Herald* of ignoring press releases and statements made by the Board and other government organizations. The Board placed full page advertisements in various papers and their managing director, Bob Wilson, was worked off his feet with radio and television interviews. Senior Board executives were forced to work late hours and weekends to cope with the crisis.

Individual journalists were subject to various forms of manipulation by Water Board public relations staff. For example, one was subject to an angry phone call during which his professional ethics were questioned and he was told that it was unlikely that any Water Board employees would want to talk to him in future. Ten minutes later the same person rang him back and in quieter tones asked him to excuse the first call but to understand that everyone in the Board was under extreme pressure. He was told that at least one employee was under doctor's orders to remain at home because of stress-related illness which was attributable to what was happening. Not surprisingly the journalist was very upset by this call and he considered what he was doing very carefully but his colleagues rallied round him and encouraged him to disregard what they saw as an effort to intimidate him.

When further fish contamination results were released to the public in July, it was at a press conference held by the Minister of the Environment. This time the Minister had an expert, a university professor, at the press conference to ensure the correct interpretations were conveyed to the media. The study had been of heavy metals in red morwong caught near the outfalls. Despite the fact that most of the fish sampled were over NH&MRC limits for mercury the Minister stated that the study showed there was no toxicological threat to humans from heavy metals discharged in effluent from ocean outfalls.[42]

The university professor compared average levels of mercury in the Sydney fish to the highest levels found in fish from Minamata Bay in Japan where more than one hundred people died and hundreds more were sick from mercury poisoning after eating the fish there. He concluded that "treated sewage as presently discharged does not constitute a hazard in terms of heavy metal accumulation."[43]

The media left the press conference with the impression that the new report gave the fish a clean bill of health. The professor's statement that one would have to eat fifty kilograms of red morwong a week continually "to get any real trouble" was shown on every television news broadcast that evening. The Minister for the Environment was even reported in the *Sydney Morning Herald* the next day as saying that the "study proved that the effluent which was being discharged from treatment plants at Malabar, Bondi and North Head was not deemed to be a health hazard for the fish."[44]

The problem was the public was not being told was that these red morwong were the very same red morwong that had been kept in a refrigerator since being analyzed for and found to be heavily contaminated with organochlorines earlier. These fish were far from being safe to eat. Fortunately, at the time I had easy access to the media and I was given a chance to point this out in the *Sydney Morning Herald* and on one of the commercial television channels. After this the professor backed away from the statements he had made about the fish being safe to eat. He was reported as saying:

> I didn't mention the organochlorines because it was not in my brief and I wouldn't talk about them anyway. . . . I made my comments on the basis that if there was no other contaminating factor, then the fish would be all right to eat. . . . Obviously if there are organochlorines I think anyone who ate the fish from there would be very foolish.[45]

OUTCOMES

The events of early 1989 came as a shock to the Water Board. The initial stories in the *Sydney Morning Herald* in January 1989 triggered other articles and stories in every media outlet in Sydney, as well as the national and international press, including *Time* magazine. People came forward with revelations about other Water Board coverups and journalists conducted their own investigations into various aspects of the story. Alan Tate and Paul Bailey at the *Sydney Morning Herald* won awards, including the prestigious Walkley award for their series of investigative articles on sewage pollution.

Many television celebrities, musicians, sporting stars, and others added their voices to the cry of outrage over the pollution. Some doctors finally spoke up. The *Sun-Herald* ran a feature on beachside doctors, more than half of whom had reported an increase in ear infections, gastroenteritis, and viral infections. Most linked these problems to beach pollution.[46] A month or so later, a group of eighty doctors, led by Peter MacDonald, called upon local councils to close fifteen northern suburbs beaches until they could be proved safe.[47] (Peter MacDonald was elected to State Parliament at the following election as an independent after campaigning strongly on the issue of sewage pollution in his electorate.)

Beach culture and seafood restaurants were an essential part of Sydney's identity but now people no longer knew whether it was safe to go swimming, surf-lifesavers threatened to walk off the job, and the fishing industry was losing an estimated half a million dollars each week as people turned away from seafood in droves. It was said that tourists were still visiting Sydney's famous beaches "but only for quick strolls rather than long days at the beach where they once splurged on ice-creams, hot dogs and souvenirs."[48] It was claimed that takings from shops and businesses at the beachside suburb of Manly were down 15 percent on the year before and some people were threatening legal actions. Property sales were also reported to be affected by the pollution publicity and some residents believed that real-estate prices were being affected.[49]

Various journalists and reporters were threatened for reporting on the pollution. Surf Reporter John Ellis (radio station 2MMM) received such a threat from a Manly businessman who claimed to represent twenty Manly businesses that had been adversely affected by Ellis's warnings to people not to swim at Manly. Kirk Willcox, long time member of STOP, lost his job as surf reporter at radio station 2JJJ at about this time, because the Australian Broadcasting Corporation said it could no longer afford surf reports. "It's ironic" Willcox said "that I've been thrown off the air at this time—when ocean pollution has finally become a front-page issue. Now I have no avenue to voice my opinion."[50]

On Good Friday 1989, almost a quarter of a million people gathered at Bondi Beach for a rock concert, the "Turn Back the Tide" Concert, staged as a protest against the pollution of Sydney's beaches. Some of Australia's leading singers and musicians donated their time and talent to the cause and the hundreds of thousands of young people, who had come along despite the occasional showers and overcast conditions, bellowed out their indignation as speakers from the stage condemned the authorities for allowing the beaches to deteriorate so badly.

To the Water Board officers, the whole episode from the *Herald*'s first revelations on 7 January to the 240,000 strong attendance at the Turn Back the Tide Concert on 24 March was just a media beat-up. After all, nothing substantial had changed from the previous year. For example, Water Board officer Leigh Richardson was reported as saying that the sudden interest in water pollution was largely a figment of the media's imagination.[51] And many others besides wondered why there was suddenly so much interest in sewage pollution. Was it just that the papers were short on stories?

However, the politicians were not so immune from public opinion and the government initiated an independent review of the Water Board proposals, although the Minister was careful to assure the Board engineers that he was not bowing to public pressure:

> Although the review has come at a time when there has been considerable public debate over the role and achievements of the Water Board and the levels of pollution of Sydney's beaches, the review is not a response to those public comments or pressures.
>
> The purpose of the review . . . is to ensure that the reputation of the Board and its employees is preserved and that the Board is not seen to be acting as "judge and jury" on matters of public controversy relating to its operations.[52]

The successful tenderer for the review, the U.S.-based engineering firm Camp Dresser and McKee, was announced in March 1989. Their local Australian affiliate, Camp Scott Furphy Pty. Ltd., have had a long association of doing work for the Board including work on treatment plants, according to the Municipal Officers Union. Camp Dresser and McKee representatives, who were working out of the offices of their Australian affiliate, told me that they disagreed with U.S. legislation that requires secondary treatment of all municipal discharges going into the ocean, because secondary treatment may only provide a small improvement over primary treatment. In Boston, they probably would not have recommended secondary treatment but it was mandatory under the legislation.

Nevertheless, Camp Dresser & McKee confirmed many of my own findings and found the Water Board's extended outfalls would not solve the sewage pollution problems in Sydney. They recommended that treatment at the main outfalls be upgraded, although they stopped short of recommending secondary treatment. They also recommended that $6 billion be spent on the Board's sewerage system over the following twenty years. In December 1989, the government announced that it would be spending more than $7 billion over the next twenty years to clean up water pollution in Sydney and surrounding areas.

CONCLUSIONS

Many people would like to excuse the Water Board engineers from responsibility for what has been happening on the beaches and in the oceans because, after all, it is the politicians and the Board members who set the budgets and it is other agencies who set the standards to be met. The engineers were only doing their job in coming up with a technological solution that would meet the required standards within the monetary constraints. Yet I found that engineers played an active role in shaping public perceptions and molding impressions. Their studies set out to justify, legitimate, and sell the technological solutions which they preferred, ones that used the ocean for sewage treatment.

For years the Caldwell Connell studies convinced politicians, other government authorities, and a whole range of laypeople that the consequences of the deepwater outfalls had been thoroughly researched and that they would work as promised. The Board was aided in this by other government experts and politicians who helped them to keep the extent of existing marine pollution secret.

It was therefore not surprising that most of the early opposition to the deepwater ocean outfalls had come from people who were philosophically opposed to the idea of ocean disposal because they felt it wasted resources, rather than people who argued that the outfalls would not clean up the pollution. Environmentalists tended to put forward reuse and recycling alternatives without challenging the claims made by the Water Board engineers and their consultants for the deepwater outfalls scheme. Because those engineers retained their credibility as experts, they were able to authoritatively dismiss the alternatives as being too costly and not feasible.

It wasn't until the credibility of the Water Board engineers had been attacked that discussion of alternatives could take place. In fact a large range of new treatment technologies, which were not previously part of the engineer's normal repertoire, emerged following the initial burst of publicity in 1989. Some of these were taken up by the Water Board, trialed, and considered for implementation, particularly those that avoided the need for biological treatment that might be sensitive to toxic materials in the sewage.

However, the Board engineers are still to be convinced that the ocean is not a suitable place for sewerage treatment. An annual environmental levy of eighty dollars per ratepayer was raised to cover the new measures that were recommended by Camp Dresser and McKee and were being demanded by the public. The Board is now planning to spend only 28 percent of the levy ($137 million) on reducing ocean pollution and has during this time paid the New South Wales government $200 million in dividends. Four years after the decision to clean up the waterways, the Board had still not decided how the treatment plants would be upgraded. The deepwater outfalls were intended to make the problem of sewage pollution less visible and now this has been achieved, I think the Board is hoping people will be happy with less treatment.

At a recent Pricing Tribunal Seminar in Sydney, Bob Wilson, General Manager of the Sydney Water Board said that the Board's main problem was the "emotionalism of the environment." The media fanfare surrounding ocean pollution was based on emotion and had distorted the picture of what the Board considered were the real problems. "Unless we get the science right" he said, "emotion can take over." What Wilson was concerned about was that the government might be swayed by public opinion to set different priorities from those that he and the experts advising him hold.

It is not the emotionalism of those wanting to protect the environment that we have to worry about, but rather the emotional attachment that some experts have to outdated ideas, professional autonomy, and status. Sewerage experts need to learn to respect community concerns for the environment and incorporate them in their designs, not dismiss them as an emotional fallacy.

NOTES

1. A. M. Rawn, "Fixed and Changing Valves in Ocean Disposal of Sewage and Wastes," E. A. Pearson, ed., *Proceedings of the First International Conference on Waste Disposal in the Marine Environment*, Pergamon Press (1959): 6–7.

2. Water Board advertisement, *The Good Weekend* (supplement to the *Sydney Morning Herald*) (12 December 1987).

3. The Ph.D. was carried out in the School of Science and Technology Studies at the University of New South Wales.

4. Joel Tarr et al., "Water and Wastes: Retrospective Assessment of Wastewater Technology in the United States 1800–1932," *Technology and Culture*, vol. 25, no. 2 (1984): 246–247.

5. Ibid.

6. See for example, T. P. Anderson Stuart, "Anniversary Address," *Royal Society of NSW*, vol. 28 (1894): 16–19.

7. For example, Henry Robinson, *Sewerage and Sewage Disposal*, London: E. & F. N. Spon (1896); Baldwin Latham, *Sanitary Engineering*, London: E. & F. N. Spon, (1878, second edition).

8. Robinson, op cit.: 45.

9. For example, The Water Board, *Clear Water. Clean Sand*, brochure; MWS&DB, *Deepwater Submarine Outfalls To Protect Sydney's Beaches*, brochure.

10. See for example, I. G. Wallis, "Ocean Currents Offshore from Sydney," *Sixth Australian Conference on Coastal & Ocean Engineering*, IEAust (1983): 210.

11. *New Scientist* (16 July 1981).

12. Caldwell Connell, *Sydney Submarine Outfall Studies*, MWS&DB (1976).

13. Ibid.: 149.

14. Ibid.: 12.

15. Fecal coliform are naturally occuring organisms found in human wastes that are used as an indicator of the presence of sewage.

16. Ibid.: 34, 149.

17. For example, Caldwell Connell, *Environmental Impact Statement, Malabar Water Pollution Control Plant*, MWS&DB (1979).

18. Caldwell Connell, *Reconnaissance Survey of Heavy Metal and Pesticide Levels in Marine Organisms in the Sydney Area* (October 1973).

19. MWS&DB, *Environmental Impact Statement, Bondi Water Pollution Control Plant* (1979).

20. Fisheries Research Institute, *Organochlorine Pesticide and Polychlorinated Byphenyl (PCB) Residues in Fish and other Aquatic Organisms in New South Wales*, Part 2, Department of Agriculture (undated).

21. I had friends in STOP and in 1989 became a member and spokesperson for STOP myself.

22. *Telegraph* (Sydney) (17 January 1977).

23. *Sydney Morning Herald* (27 February 1989).

24. *Engineers Australia* (21 April 1989): 4.

25. Sharon Beder, *Toxic Fish and Sewer Surfing*, Sydney: Allen and Unwin (1989).

26. A. J. Priestley, Review, *Search*, vol. 21, no. 3 (April/May 1990).

27. Alan Tate and Paul Bailey, "Fish off Sydney Beaches Polluted," *Sydney Morning Herald* (7 January 1989): 1.

28. SPCC, Annual Report 1987–88: 30.

29. Bioaccumulation of Organochlorine Pesticides Near the Malabar Ocean Outfall, Meeting Notes (18 May 1988).

30. Letter from Australian Underwater Federation to W. Forrest, Deputy Director, SPCC.

31. Dietrich Georg, "Engineers Criticised for not Going Public on Pollution," *Engineers Australia* (26 January 1990): 16.

32. For details of these, see Beder, op. cit., chapter 3.

33. Water Board advertisement, *The Good Weekend* (supplement to the *Sydney Morning Herald*) (12 December 1987).

34. Cover note on early copies of 1987 Bio-Accumulation Report.

35. For example, Bob Wilson, Managing Director of Water Board, Sewage Summit, Bondi Pavillion (18 February 1989).

36. Letter from Minister for Agriculture and Rural Affairs, Ian Armstrong, to Minister for the Environment, Tim Moore (13 July 1989).

37. Letter from D. W. Connell, Griffith University, to Peter Fagan, Water Board (4 September 1989).

38. Letter from Jack Anderson, Southern California Coastal Water Research Project, to Tony Misckiewicz, Water Board (24 July 1989).

39. Jack Anderson, "Overview of the Planned Environmental Monitoring Programme," (1989).

40. *Sydney Morning Herald* (17 January 1989).

41. Ibid. (8 March 1989).

42. Press Release from Minster for the Environment, Tim Moore (3 July 1989).

43. Evening news, all channels (3 July 1989).

44. *Sydney Morning Herald* (4 July 1989).

45. Ibid. (6 July 1989).

46. *Sun-Herald* (12 March 1989).

47. Ibid. (16 April 1989).

48. *Sydney Morning Herald* (28 January 1989).

49. *Sun-Herald* (16 April 1989).

50. *Sydney Morning Herald* (26 January 1989).

51. Ibid. (28 January 1989).

52. *Aquarian* (April 1989).

3

Fluoridation: Breaking the Silence Barrier

INTRODUCTION TO THE ISSUE SAID TO BE "BEYOND SCIENTIFIC DEBATE"

The scene is the town hall of Moruya, a small town on the south coast of New South Wales, Australia. I am sitting on the stage, waiting my turn to speak at a public meeting called to discuss whether the town water supply should be fluoridated. At the speaker's lectern is one of the leading fluoridation campaigners of the Australian Dental Association. He is telling the audience that some water supplies contain fluoride naturally, which is true, and therefore that fluoridation must be safe. This does not follow logically, nor is it true. Will the audience understand and believe me, an independent scientist opposing the dental and medical establishment, when I present evidence that fluoridation is harmful to human health and that its benefits have been exaggerated? The dentist is now claiming that, if fluoridation were harmful, the human race would have been already wiped out by natural fluoride. My determination strengthens—in the face of such ignorance or deceit, I will not give up.

In most English-speaking countries, the fluoridation of water supplies is presented by dentists, doctors, and public health officials as the cornerstone of dental public health. In such countries it has been endorsed by the dental and medical associations and departments of health. It is described as having enormous benefits but no risks, and even as being "beyond scientific debate."

Fluoridation is the addition of fluoride to drinking water to increase the natural fluoride content to a concentration of about one part per million (1 ppm), that is, 1 milligram of fluoride per liter of water. Although there are some regions of the world where natural fluoride exists in drinking water at concentrations of 1 ppm or higher, in the vast majority of water supplies the natural fluoride concentrations are typically one-tenth to one-fifth of 1 ppm, and so fluoridation generally leads to considerable increases in people's intake of fluoride.

The purpose of fluoridation is to reduce the prevalence of tooth decay, called 'dental caries' in the dental, medical, and public health literature. Unlike chlorination, which is designed to kill bacteria, thus making water safer to drink, fluoridation is designed to treat people, and so may be considered to be mass medication. This is an important ethical objection to fluoridation. Furthermore, some opponents describe fluoridation as compulsory medication. More accurately, I would say that it is medication which is expensive to avoid, since people who do not wish to be dosed have to purchase bottled water or equipment to remove fluoride from their drinking water.

Apart from the ethical issues are the political issues of who controls, funds, and profits from fluoridation, and the scientific issues of the determination of the dental benefits, health hazards, and environmental impacts of fluoridation. As a research scientist, I have concentrated on the scientific issues, while taking an interest in the ethical and political contexts.

The practice of fluoridating drinking water supplies began in the United States in the 1950s, and then spread to Canada, Australia, New Zealand, Ireland, and a few other countries. But, fluoridation is almost nonexistent in western continental Europe or in most other non-English-speaking countries.[1] It has been discontinued in Sweden, Holland, Germany, and Finland, mainly on account of concerns about its health hazards, known or potential. Only a few percent of the world's population drink artificially fluoridated water, although that information is rarely revealed to the peoples of heavily fluoridated countries.

Although the establishment 'experts' generally receive better coverage in the media than 'dissidents' on most environmental, health and political issues, only in the case of fluoridation have the 'experts' succeeded in convincing the vast majority of people in whole countries that opponents must be either cranks, extreme right-wingers or health 'faddists.' This remarkable propaganda success has been achieved primarily by trading on the authority of the medical profession and by putting pressure on 'dissident' medical doctors, dentists, and scientists to keep silent. The stereotyping of opponents has placed pressure on scientific and professional journals and the media not to publish material critical of fluoridation. So, the need to break the silence barrier is a special feature of opposition to fluoridation.

In this chapter, I explain how I became involved in the issue, how I found internal contradictions and misrepresentations in the pro-fluoridation case, how I campaigned against fluoridation, how the establishment 'experts' tried to suppress me, and conclude by offering some lessons. Boxes are included on (1) the fluoridation power structure and (2) how fluoride acts on teeth. There is also an Appendix summarizing my critique of fluoridation.

HOW I BECAME INVOLVED

As one of the offspring of an engineer and a poet, I could be expected to draw upon both the disciplinary and holistic approaches to problem solving. So it will hardly be surprising that I became a research scientist with broad interests and concerns: social justice, environmental protection, and the health hazards of environmental chemicals and ionising radiation.

Although my Ph.D. research was mostly on a specialized topic in applied mathematics, my subsequent research spanned a wide range of practical applications of mathematics and other natural sciences. As a postdoctoral researcher at Imperial College, London, United Kingdom, I performed analysis of ground and satellite data in space science. Then, as a research fellow and lecturer at the Australian National University, I collaborated with neurobiologists on mechanisms of insect smell and vision, and also initiated my own research on cooperative effects in biological catalysts which change their shape. In the CSIRO[2] Division of Mathematics and Statistics from 1975 to 1985, I worked on generation planning in electricity grids and the economic value of wind electric power, among other things.

This breadth of experience has been of great of value to me in taking on interdisciplinary public issues such as fluoridation. My involvement in public issues was stimulated in part by the shocked discovery that my Ph.D. thesis had been used by hydrogen bomb scientists. This experience, imposed on my scientific training and interdisciplinary inclinations, led me naturally into issues of science and society from 1969 onwards.[3]

For most of the 1970s, I was either vice president or secretary of the Society for Social Responsibility in Science (SSRS) in Canberra, Australia. Over that period SSRS had about two hundred members, mostly scientists and academics, and aimed to inform decision makers, scientists, and the public about the social consequences and implications of science and technology. As secretary, I had an overview of almost all its activities, which were mostly on environmental issues, and was also able to introduce some of my own particular areas of interest—a critique of modern medicine,[4] support for the new public health and community health movements[5] and energy alternatives.[6]

So it was not surprising that, when SSRS occasionally received letters from people who believed that they suffered ill-effects from drinking fluoridated water, I was ready to investigate the issue further.

SEARCHING THE SCIENTIFIC LITERATURE

As a research scientist, it was natural for me to begin, in the mid-1970s, with a thorough review of the scientific literature on the alleged benefits and health hazards of fluoridation. Also, because dental and medical proponents claimed a scientific basis for fluoridation, I felt that I had to go back to the original papers in dental, medical, and scientific journals, and not allow myself to be restricted to official reviews and reports of inquiries.

The basic pro-fluoridation position was easy to identify. In extensively fluoridated countries there are many official leaflets, brochures, and reports spreading the message that fluoridation produces enormous reductions in tooth decay and is completely safe. In Australia, such documents are produced mainly by the Australian Dental Association (ADA), the National Health and Medical Research Council (NH&MRC), and the state departments of health. But, in the 1970s, few official documents contained references to medical and scientific papers attempting to justify the claims of safety.

On the alleged dental benefits, the pro-fluoridation reviews did refer to the early studies of tooth decay in naturally fluoridated regions of the United States by H. T. Dean, the "father of fluoridation," and others. They also took as part of their foundations the early trials of artificial fluoridation which commenced in several North American cities in the mid-1940s.

When I read the original papers, I was amazed at the arbitrary selection of data and the absence of statistical analysis. The scientific standard of many of the 'classic' papers was that of junior high school rather than university research. Nevertheless, the sheer quantity of papers reporting enormous benefits from fluoridation, natural or artificial, suggested to me initially that the results might be genuine. In the 1980s, new evidence on the decline of tooth decay in unfluoridated areas and the mechanism of action of fluoride on teeth brought me to reconsider that position (see below).[7]

The task of finding original medical and scientific literature on the health hazards of fluoridation was made difficult by pro-fluoridationists' claims that such evidence did not exist. Their leaflets and reports claimed that someone would have to drink a bathtub full of fluoridated water to suffer ill-effects. I found this to be misleading, because it confused the acute effect of a single high dose with the chronic effects of drinking small doses over years and decades. When fluoridated water is drunk, about half the fluoride is excreted by the kidneys (provided they are working properly) and the rest is stored in the bones, accumulating until death. It is now widely accepted that the bones become heavier, but more brittle. Over a normal lifetime, people living in fluoridated areas can store much more fluoride in their bones than that dissolved in a bathtub of fluoridated water.

In searching the literature on the hazards of fluoridation to bones and other organs, I was helped by the books and unpublished reports of the anti-fluoridation movement which contained many useful references. But, I had to examine their information critically too, because some parts of the grassroots anti-fluoridation movement are bound by their own traditions.[8] But, I soon found several scholarly papers presenting evidence that skeletal fluorosis, a disease of the bones and joints, is endemic in several naturally fluoridated areas of the world.

Skeletal fluorosis is similar in symptoms to arthritis. Like arthritis, it can become crippling in some cases. In naturally fluoridated areas of India and several other countries, skeletal fluorosis is a well-recognized public health problem, particularly for the aged. In India it is even observed in some villages where the fluoride concentration is as low as 0.7 parts per million.[9] Yet, when proponents of fluoridation are asked about skeletal fluorosis, they often create the false impression that it is only seen when fluoride concentrations in drinking water are much higher, 8 ppm or more.[10] When confronted with the studies of skeletal fluorosis at 0.7 to 2 ppm, they either deny them or attempt to label these as special or peculiar cases.

Several other papers I found were by medical doctors and dentists who reported intolerance or hypersensitivity reactions to artificially fluoridated drinking water and fluoride tablets. The reactions include skin rashes, stomach pains, and effects on the nervous system. Clinical reports of these reactions have been checked by 'blind' tests, in which the patients did not know when they were ingesting fluoride and when they were ingesting a placebo. There has been no properly designed large-scale epidemiological study on such reactions. However, a pilot study in the United States indicated that possibly about 1 percent of the population might be sufferers.[11]

In the professional dental literature I found it well recognized that the ingestion of fluoride during early childhood can damage the enamel-forming cells, and that this in turn produces the particular type of dental mottling known as dental fluorosis.[12] But, although its occurrence is clear evidence of physiological damage, most proponents of fluoridation describe dental fluorosis as merely a 'cosmetic' effect. To me this seems like shrewd marketing rather than an open acknowledgement of well-established disease.

At this stage of my research it was obvious that the official pro-fluoridation reports and leaflets had ignored important scientific/medical papers which raised doubts about the alleged safety of fluoridation, or dismissed them on ludicrous grounds, or misrepresented them. My appetite for the fluoridation issue was whetted by these discrepancies and I decided to devote some time to fluoridation as a serious issue of public interest science.

My determination to do something about it was strengthened by reading the report of the Tasmanian Royal Commission,[13] which in parts verged on racism. It did discuss skeletal fluorosis, but denigrated the overseas evidence by classifying the disease as occurring in 'native' populations and therefore by implication as being irrelevant to (white) Australians. As in the case of the issues of nuclear energy and the health hazards of ionizing radiation,[14] I found that the establishment 'experts' on fluoridation were misleading the public and decision makers.

FINDING ALLIES

By writing to or phoning leaders of the anti-fluoridation movement in Australia, I was put in touch with other scientists, dentists, and medical doctors here and overseas who had doubts about the safety and/or the effectiveness of fluoridation.

In the 1970s and early 1980s, my main professional and scientific advisers on fluoridation were Dr. Philip R. N. Sutton, a retired dental researcher and senior lecturer from the School of Dentistry, University of Melbourne,[15] Albert Burgstahler, Professor of Chemistry at the University of Kansas, and Mr. Glen Walker, a retired businessman with expertise in metal finishing and electrochemistry, who was and is still the coordinator of the grassroots antifluoridation movement in Australia.[16]

From the mid-1980s onwards, I benefitted greatly from regular correspondence with Dr. John Colquhoun of Auckland, New Zealand, who was formerly chairperson of the Fluoridation Promotion Committee of New Zealand and is now a leading opponent on the world scene. From the late 1980s, I corresponded with Dr. John R. Lee, a Californian medical doctor. These and other antiestablishment 'experts' exchange information and test their ideas in a fruitful way. Between us, we span a wide array of dental, medical and scientific knowledge and experience.

I browsed regularly in dental and medical libraries and identified the key journals which publish papers on fluoridation. With a little help from my medical and dental mentors, I soon learned the basic jargon and found that professionals sometimes make damaging admissions in their own journals which they would never dream of making to the public. Subsequently it turned out to be valuable to be able to quote these admissions in my publications on fluoridation and in the rare public debates.

MY FIRST PUBLICATIONS ON FLUORIDATION

By the mid-1970s I had reached the stage where I wished to publish the evidence in support of my concerns about fluoridation. But, in the climate where I would immediately be labeled as a crank, fanatic, or faddist if I raised the issue, I could find few outlets apart from local newspapers and radio in the towns where controversy about fluoridation was raging.

Meanwhile, my main voluntary work for SSRS was conceiving and then editing a book called *The Magic Bullet*, a critique of modern medicine, something which was new to Australia at that time.[17] The chapter on "Environment and health," written by the eminent human ecologist, Dr. Stephen Boyden, and myself, referred to fluoridation as an example of an 'antidotal' form of preventive medicine, rather than a 'corrective' form like having adequate vitamin C in the diet to prevent scurvy. Fluoridation is 'antidotal,' like a dental fissure sealant, because, contrary to much pro-fluoridation propaganda, dietary fluoride in doses of typically a milligram per day is not necessary for sound teeth. Some people have excellent teeth yet have fluoride intakes far below the level recommended by pro-fluoridationists.

The Magic Bullet created widespread public and media interest[18] and sold out rapidly. As a follow-up, I became the co-organizer of a national conference on *The Impact of Environment and Lifestyle on Human Health*.[19] The conference was devoted to reducing the power of the medical profession over health issues, which are nowadays mostly environmental and lifestyle in origin, and enhancing the role of public and community health. The time was ripe for such a conference, which turned out to be a great success.

At the conference I took a risk and presented a paper entitled "A closer look at prevention," in which I included fluoridation as the principal example of a form of preventive medicine which may have health hazards.[20] Possibly because the paper was presented humorously in an appropriate context and was not simply a head-on attack on fluoridation, it was well received. Perhaps for the first time in Australia, a paper reviewing some of the health hazards of fluoridation was presented to an audience of public health professionals, medical doctors with concerns about environment and lifestyle, other health professionals and academics.

Encouraged by these limited successes in breaking the professional silence barrier, I then wrote a critical review of the 1976 pro-fluoridation report by the British Royal College of Physicians.[21] Although this was a direct attack on fluoridation, my newly established credibility in the public/community health field apparently enabled the paper to receive serious consideration by *Community Health Studies*, journal of the Australian Public Health Association. After I had responded to the comments of a referee who accused me of bias, the journal published my paper.[22]

A CONTROVERSIAL DEBATE GAINS MEDIA COVERAGE

In 1979, a visit to Australia by the U.S. biochemist, Dr. John Yiamouyiannis, principal author of a paper claiming that there is a link between fluoridation and cancer,[23] offered the opportunity to air this controversial issue more thoroughly. Almost as soon as he arrived, the medical and dental establishment attacked Yiamouyiannis personally in the media, but seemed unwilling to debate the scientific evidence he put forward. So I arranged for SSRS to sponsor a scientific debate at the Australian National University between Yiamouyiannis and a spokesperson for the NH&MRC. The NH&MRC first took the traditional pro-fluoridation stance that the subject was beyond scientific debate, but I had managed to interest the *Canberra Times* in the issue and the NH&MRC had placed itself publicly in a position where it either had to put up or retract. So, reluctantly, they nominated a speaker, retired professor of pharmacy Roland Thorp.

In the debate which followed, it soon became obvious that Thorp had little specific knowledge of the data on fluoridation and cancer. He simply gave the standard general pro-fluoridation speech. He was unable to answer Yiamouyiannis' specific points on fluoridation and cancer, and could not or would not reveal who in Australia had assessed the scientific literature on fluoridation and cancer for the NH&MRC and had pronounced Yiamouyiannis wrong. The debate was reported fairly in the *Canberra Times* and subsequently there was some interesting correspondence.

It must be stressed that at no time did SSRS or I take the position that fluoridation causes cancer. In my view, there is conflicting scientific evidence, but sufficient grounds for concern to require further studies and for SSRS to provide a public forum for debate.[24]

A response of the medical-dental establishment was to wait until I was overseas, giving a paper at an international conference on wind energy.[25] In my absence a group of dentists and doctors met with my fellow SSRS committee members to pressure SSRS to drop the issue. Also the proponents of fluoridation held a joint workshop on fluoridation with the Australian Statistical Society, at which only proponents were speakers. The pro-fluoridationists clearly needed the support of statisticians to refute the alleged fluoride-cancer link. Disappointed at the lack of support of my colleagues in SSRS on this and other issues, I resigned as secretary and redirected my energies into other community groups.

BOX 1. The Fluoridation Power Structure: Its History and Tactics

In Australia, the principal institutional proponents of fluoridation are the National Health and Medical Research Council (NH&MRC), which first endorsed fluoridation in 1952, the Australian Dental Association (ADA), the Australian Medical Association (AMA) and the State Departments of Health. Until a few years ago, the Federal Department of Health also played an important role, but then its dental health branch was closed as part of a cost-cutting program. Martin's book[26] lists some of the main personalities in the Australian fluoridation debate and surveys their views.

Within the above pro-fluoridation organisations, very few people seem to have read the original scientific, medical, and dental literature on fluoridation and very few can stand up on a public platform or at a university and credibly debate the issue with a scientific opponent who has. Their support for fluoridation is based on simplistic teaching in dental and medical schools, the endorsement of fluoridation by the executive committees of their professional associations, and propaganda produced for decision makers and the public by a small group of pro-fluoridation cadres. This is a reflection of the way fluoridation has been promoted and implemented—by lobbying and capturing the support of a few top people in key institutions. It has never been a grassroots movement. When the public has been given the opportunity to express an opinion about fluoridation—in referenda, public debates, letters to newspapers and petitions—the majority usually opposes it.

In the United States, as late as 1943, fluoride was officially regarded as a pollutant of air and water, and the U.S. Public Health Service (USPHS) regarded fluoride concentrations in excess of 1 ppm as constituting grounds for the rejection of drinking water supplies. But, research funded by the aluminium industry, for which the disposal of fluoride used in the smelting process was an expensive problem, suggested that fluoride may be required for tooth formation. Then a group of dentists and state dental health officials in Wisconsin carried out a long lobbying campaign. Eventually, in 1950, they succeeded in getting the USPHS to reverse its previous cautious stance and endorse fluoridation. Like the NH&MRC in Australia, the USPHS exercised enormous influence through its funding of research grants. From the endorsement by the executive committees of the USPHS, NH&MRC and the medical and dental associations many others flowed.[27] Although many medical doctors and a few dentists spoke out against fluoridation at the time,[28] they were not organized and their objections were overridden by the rising tide of official endorsements.

As fluoridation spread in the United States, an eminent allergist, Dr. George L. Waldbott, reported that some of his patients suffered allergic, intolerance, or hypersensitivity reactions from fluoridated drinking water. His books also reveal the unprofessional means used by some members of the fluoridation establishment to try and discredit him and to keep his reports out of medical and scientific journals and out of the media.[29] Indeed, when the U.S.-based medical journals would no longer publish papers on the health hazards of fluoridation, Waldbott even had to publish one of his papers in the *Medical Journal of Australia*.[30] Subsequently, the curtain of silence fell also in Australia. Reading Waldbott's 1965 book helped to prepare me for the similar attempts at intellectual suppression to be used against me in the 1980s.

By using the authority of the medical, dental and public health estab-lishments, the proponents of fluoridation succeeded until the mid-1980s in Australia and New Zealand in keeping the scientific evidence against fluo-ridation out of almost all mainstream media.[31] The fluoridation establish-ment brought pressure to bear from the highest levels on editors and pub-lishers of newspapers, magazines and books and on producers of programs in the electronic media. After the unexpected broadcast in 1979 of an Australian Broadcasting Commission '4 Corners' television program, which presented both sides of the fluoridation issue, senior medical doctors and dentists influenced the ABC to keep the subject off the air for years after-wards.[32] A journalist on a leading Australian newspaper published in Melbourne, *The Age*, told me that he had been instructed to drop the issue or be fired. Chris Wheeler, the editor of an Auckland, New Zealand, subur-ban newspaper, the *Shore News*, was fired on the day in 1988 when he brought out an issue containing a large number of letters-to-the-editor about fluoridation from both sides.

In the United States, the history of settlement by dissenting religious communities and the tradition of local democracy allowed local communi-ties a greater say in decision making and may have helped keep the propor-tion of people with fluoridated drinking water down to 50 percent. But, in Australia, with its authoritarian forms of state government descending from the colonial governments of convict settlements, legislation promoting flu-oridation is distinctly antidemocratic. For example,

- In the State of New South Wales, the Fluoridation of Water Supplies (Amendment) Bill 1989 has the effect of preventing local governments from terminating fluoridation.
- The Victorian Health (Fluoridation) Act 1973 allows the State Government to impose substantial daily fines on water authorities which, following the will of communities which elect them, decline the govern-ment's request to fluoridate.
- In the State of Tasmania, clause 13 of the Fluoridation Act 1968 makes it illegal for local governments to hold polls to determine public opinion on fluoridation.[33]

THE ANZAAS SYMPOSIUM GAINS WIDE PUBLICITY

In the early 1980s, it was very difficult to gain open discussion of the health hazards of fluoridation in the mainstream media. However, by addressing public meetings, speaking on local radio, and writing letters to local newspapers, I did help several local com-munities to fend off attempts by the New South Wales Government to impose fluoridation upon them. I was spending most of my time on windpower research and on building up the Australasian Wind

Energy Association of which I had been a co-founder in 1980. But I still kept up an occasional watching brief on the dental literature on fluoridation.

My own research on fluoridation was reactivated by the publication of papers overseas reporting that there had been large declines in tooth decay over the 1960s and 1970s in several *unfluoridated* developed countries.[34] I was also aware of evidence of similar declines in Australia—in prefluoridation Sydney and unfluoridated Brisbane.[35] These declines had commenced too early to have been caused by fluoride toothpaste and there was evidence suggesting that fluoride tablets had not played a major role. The obvious question, avoided by the dental researchers and fluoridation promoters, was: if similar large reductions in tooth decay were occurring over a similar period in both fluoridated and unfluoridated areas, is it not likely that the same factor was responsible in both cases? If so, that common factor could not be fluoridation.

In Australia, the promoters of fluoridation had not revealed in their official reports[36] even a hint of the new scientific evidence. I thought that the new material would be of interest to the Australian scientific community and also possibly to the media. So I enlisted the collaboration of Dr. Philip Sutton, and together we convened a symposium on fluoridation at the 1985 Festival of Science sponsored by the Australian and New Zealand Association for the Advancement of Science (ANZAAS). We invited Wendy Varney, who had just written an insightful political science thesis on fluoridation,[37] to join us as a speaker and then, to liven things up even further and to inject 'balance,' we invited the pro-fluoridation Australian Dental Association (ADA) and NH&MRC to each provide a speaker as well.

The ADA wrote back promptly, not to us, but to the organizers of the ANZAAS Festival of Science, declining to participate and questioning our motivations. Some of the ANZAAS organisers interpreted this letter as an unsubtle attempt to stop the symposium. The NH&MRC only replied about a fortnight before the symposium, stating that they would only participate under conditions which were by then essentially impossible to fulfill.

Fortunately, these establishment responses failed to stop the symposium. Indeed, when we explained the situation to the media, they found it to be 'news' and gave excellent advance publicity for the symposium. As a result, about one hundred people attended, including the media and many scholars who were previously uncommitted on this issue. For the first time, widespread media publicity was obtained in Australia for the evidence that the benefits of fluoridation have been greatly exaggerated, that there are genuine health hazards from fluoridated water, and that the promotion of fluoridation and fluoride products has been funded in part by vested interests such as the aluminium and sugary food industries.

In the subsequent media coverage, the fluoridation proponents were forced to come out and debate. Unaccustomed to discussing openly the issue which they had labelled as 'beyond scientific debate,' they did not offer meaningful answers to many of the points raised at the symposium by Philip Sutton, Wendy Varney, and myself, but instead they tried to disparage us personally. In participating in this symposium and in the media reports, I was described as a CSIRO scientist, as I was entitled, but I was careful to state that my conclusions were not necessarily those of any organization with which I was associated.

The real counterattack by the proponents of fluoridation took place behind the scenes. The ADA wrote to the chairman of my employer, CSIRO, and the Minister for Science and Technology, who is responsible for CSIRO, complaining about my 'activities,' describing them as "misleading, verging on fraudulent" and attacking me for allowing myself to be identified as a CSIRO scientist.[38] Fortunately, neither the Chairman nor the Minister was impressed with these heavy-handed tactics. A CSIRO administrator informed me about the complaints and I was then able to obtain the correspondence under Freedom of Information. The Minister, Barry O. Jones, had annotated one ADA letter with the following comment: "Had the possibility of countering his argument occurred to their collective minds? . . . Perhaps unfamiliar with the concept of scientific debate."

Dentists and medical doctors are more vulnerable to this kind of pressure than I was. Several cases of intellectual suppression of dentists, scientists, and medical doctors concerned about fluoridation are described by Waldbott,[39] Moolenburgh,[40] and Martin.[41]

BOX 2. How Fluoride Acts on Teeth

In the early days of fluoridation, the 1950s and 1960s, dental researchers believed that fluoride had to be swallowed to be effective. The theory was that fluoride acts systemically (i.e., internally), going from the bloodstream into the tooth enamel, allegedly strengthening the teeth. But measurements showed that hardly any fluoride goes back from the bloodstream into saliva. About half the ingested fluoride is stored in the bones where it builds up over a lifetime; the rest is excreted in urine by the kidneys, provided they are functioning properly. Furthermore, the systemic theory did not explain the action of fluoride toothpastes and gels, which became widely used in the late 1970s, requiring a mechanism based on the action of fluoride on the surfaces of teeth. So then, dentists believed in a mixture of mechanisms with both systemic and surface action.

But in the 1980s, researchers observed that, contrary to the systemic theory, the amount of tooth decay in individuals' teeth does not seem to depend on the fluoride content of their dental enamel and that the observed differences in fluoride level in dental enamel between fluoridated and unfluoridated areas were too small to explain large differences in tooth decay. Moreover, experiments on laboratory rats showed that, when fluoride was released gradually into the bloodstream without first passing over the teeth, there was no reduction in tooth decay, but if the fluoride at high concentrations was released in the mouth, there was a reduction.[42]

So, there is now a large body of scientific evidence indicating that there is little or no benefit from swallowing fluoride. Rather, fluoride seems to work by its surface action on the teeth.[43] Some establishment experts, such as Professor Ole Fejerskov in Denmark, accept this while others, perhaps recognizing the damage this admission does to the case for the fluoridation of *drinking* water, still ignore the scientific evidence and maintain that systemic and surface actions are about equally important.

PUBLICATION IN *NATURE*

Following the success of the ANZAAS symposium, I felt that it was time to foster an international scientific debate on the alleged enormous benefits of fluoridation. So I assembled all the data I could find on the decline in tooth decay in unfluoridated areas, summarized it in a form comprehensible to scientists who are not dentists, incorporated new data from the Australian School Dental Services, posed the 'obvious question' about the mechanism of the decline in tooth decay in unfluoridated areas, offered some possible answers, gave the paper a catchy title, and submitted it to what is arguably the leading general science journal in the world, *Nature*.

A few months later, the editor of *Nature* sent back a referee's report which presented the usual profluoridation line. I pointed out to the editor that my original manuscript had already answered most of the referee's criticisms. To account for the remaining points I made some minor revisions and resubmitted the paper. To my delight,"The mystery of declining tooth decay" was published in July 1986.[44] I think it must have helped my credibility as a serious scientist with the editor that over the previous sixteen years I had already published several refereed research papers in his journal on such 'hard science' topics as astrophysics, space physics, and wind-power.

The publication of such a substantial, controversial paper in *Nature* gained media coverage around the world. It was a major breakthrough for the anti-fluoridation case. It also strengthened my links with overseas scientists, dentists, and medical doctors who were questioning fluoridation, including Albert W. Burgstahler from the United States and John Colquhoun from New Zealand.

The counterattack of the fluoridation establishment was to cir-culate covert critiques misrepresenting my paper, to spread the false statement that my paper had not been refereed,[45] and to put pres-sure on the editor of *Nature* which could have stopped him publish-ing any further articles by me on fluoridation.

I only learned of the last move several years later when some-one in the United States sent me a copy of a letter and an attached unpublished critique of my *Nature* paper, which had been addressed to the editor of *Nature* by one of Australia's most vocal pro-fluori-dation campaigners of the 1980s, Dr. Graham Craig. Contrary to the normal scientific practice of encouraging open debate, the letter (dated 15 August 1986) commenced: "This letter and its enclosures are not intended for your correspondence columns." I had not previ-ously seen this material, although the way in which it reached me suggested that it must have been circulated widely around the world. Craig's material is very easy to refute, so it does not surprise me that it was not submitted for publication.

Someone also sent me a copy of a letter, dated 18 September 1986, from the then head of Dental Health in the Federal Department of Health, Dr. Lloyd Carr, to Dr. David E. Barmes, Chief of Oral Health, World Health Organisation. Carr's letter was obvi-ously a response to a request to "Please explain and counter the Australian data used in Diesendorf's *Nature* paper."[46] There could be no doubt that the publication of my *Nature* paper had upset the international fluoridation establishment.

CAMPAIGNING FROM THE
AUSTRALIAN INSTITUTE OF HEALTH

My appointment in 1988, to the position of senior research fellow at the Australian Institute of Health (AIH), the Australian government's health statistics institute, gave me opportunities to create further discussion of fluoridation in scholarly and public health circles. My main work at AIH was to analyze data on the use and costs of medical services in Australia. During the job interview it was made clear to me that I would not be permitted to do research on fluoridation. Fortunately, I had just completed a period of research as a Visiting Fellow at the Australian National University, where I had examined critically some of the well-known studies done in Australia and overseas which purported to prove enormous dental benefits for fluoridation.[47] I had found that these 'classic' studies were so poorly designed that they were almost worthless. Upon joining AIH my immediate unofficial goal was to publicize this latest work rather than to do further research on fluoridation.

So, I gave two seminars on fluoridation, which were well received by all except the medical and dental establishment. The proponents of fluoridation try very hard to diminish the credibility of anti-fluoridation speakers, so it must have been galling for them to see me identified as an AIH researcher at these seminars. Immediately after the second seminar, the Director of AIH suggested that I keep silent about fluoridation in future, but I did not take this advice.

Also in 1988, I was invited to Brazil to take part in an international scientific symposium-debate on fluoridation, with several scientists or professionals on each side. The audience consisted of water supply and environmental engineers, dentists, medical doctors, and public health officials. This was a valuable experience, both in testing my arguments against some of the world's leading pro-fluoridationists and in being part of a team with top-notch anti-fluoridationists, such as Dr. John Colquhoun and Dr. John R. Lee. On the other side, I was impressed with the manner of presentation of the American pro-fluoridation dentist, Dr. Herschel S. Horowitz, who was a dramatic speaker with professionally prepared slides, but I could see that he was limited by the poor content of the pro-fluoridation case. Despite our hand-drawn slides, we must have communicated to the audience the logic and conviction of our case, because an outcome of the symposium was that the proposed expansion of fluoridation in Brazil was stopped.[48]

In 1989, I took some leave from AIH and went on a round-the-world lecture tour, speaking on fluoridation at the University of Sheffield, United Kingdom; Dunn Nutrition Laboratory in Cambridge, United Kingdom; St Thomas's Hospital in London, United Kingdom; the New York State Health Department Inquiry;

the U.S. Environment Protection Agency in Washington, D.C.; and Stanford University in California. This trip contributed to breaking the silence barrier at some eminent institutions and also gained some limited media coverage for the anti-fluoridation case in these 'difficult' countries.

Back in Canberra, I gave evidence before a local government inquiry into fluoridation. The committee was divided and eventually accepted a compromise proposal made by another witness, Professor Bob Douglas, head of the National Centre for Epidemiology and Population Health. The committee recommended that the fluoride concentration in Canberra's drinking water be halved, and this was eventually implemented. But the ADA and AMA lobbied the local government and opposition parties, with the result that, following a change of government, the fluoride level was restored to 1 ppm in early 1992.

Subsequently, some of the lobbying material used by the ADA was published as an anonymous article in the *ADA News Bulletin*. The article contained a series of falsehoods about and misrepresentations of my work and that of John Colquhoun that were so gross that they were defamatory, according to legal advice received.[49] As a consequence, both Dr. Colquhoun and I managed to get our replies exposing the misrepresentations published in full in *ADA News Bulletin*.[50] But that did not restore the fluoride level in Canberra's water supply to the less harmful level of 0.5 ppm.

THE NH&MRC INQUIRY

In 1989, in response to a joint letter by John Colquhoun, Philip Sutton, and myself, the NH&MRC set up a new Working Group to hold an inquiry into fluoridation and into our allegations of misrepresentations and misuses of scientific data by some fluoridation proponents.[51] On the surface, the final report,[52] which appeared in 1991, is a whitewash of fluoridation and its leading proponents.

For instance, the Executive Summary contains such misleading statements as:

> The Working Group could find no evidence *within Australia* of skeletal fluorosis. . . . There is no evidence of adverse health effects attributable to fluoride *in communities exposed to a combination of fluoridated water (1 ppm) and contemporary discretionary sources of fluoride* (italics added).

The phrases in italics exclude the well-founded overseas evidence of skeletal fluorosis, which was acknowledged cautiously in the main body of the report, but most people reading only the Executive Summary would not realise this. The result is that most readers are led to assume incorrectly that there is no evidence of adverse health effects attributable to artificially or naturally fluoridated water. The pro-fluoridation bias of the report is also demonstrated by its failure to cite in its extensive bibliography the relevant published scholarly papers on fluoridation of Dr. Colquhoun, Dr. Sutton, and myself.[53]

But clearly the Working Group was nervous about some of the scientific evidence we had presented and must have felt that they had to cover themselves. So, the fine print of the report admits cautiously that:

- some 'isolated' cases of skeletal fluorosis are observed in some places overseas where the fluoride concentration in drinking water is as low as 0.7 ppm;
- there is 'an urgent need' to monitor the levels of fluoride exposure and dental fluorosis in Australia;
- some infants and children are overdosed with fluoride;[54]
- the quality of the early intervention trials to determine the benefits of fluoridation 'was generally poor . . .'

Neither our submission nor the NH&MRC report considered the recent revelations that there are more hip fractures (often fatal) in elderly women in fluoridated areas of the United States and Britain than in unfluoridated areas. Much of that evidence was published during the course of the NH&MRC inquiry.[55]

Although the NH&MRC report stated that the Working Group "found no evidence of fraud or misleading presentations of data," we have published the evidence for anyone to see.[56] After the NH&MRC inquiry one of the leading old guard fluoridation proponents, Dr. Graham Craig, suddenly left Sydney University and the battlefield, and several other members of the working party responsible for the misleading 1985 NH&MRC report have subsequently retired from the scene.

Professor Tony McMichael, the epidemiologist who chaired the new Working Group, and Professor A. J. Spencer, a dentist/statistician member of the Working Group, seem to have become leaders of a new guard for fluoridation. Although I consider them to be more sophisticated scholars than many of the old guard, I am not impressed with some of their tactics. For instance, as principal author of a laudatory review of the NH&MRC (1991) report, written in the form of an editorial in the *Australian Journal of Public Health*, McMichael failed to declare his role as chairperson of the Working Group. Furthermore, the 'review' misrepresented the work of John Colquhoun and myself, and even misrepresented some of the conclusions of the author's own report, making them appear more pro-fluoridation than they are. Fortunately, the journal published our replies.[57]

In early 1990, my submissions to the NH&MRC inquiry, revised and updated, were published as two major review papers on the alleged benefits and health hazards of fluoridation.[58] The main points from these papers, together with the ethical and political dimensions of the fluoridation issue, are listed in the Appendix.

Shortly after the publication of these papers, I resigned from the Australian Institute of Health to became coordinator of the Australian Conservation Foundation's Global Change Program, a national campaign to reduce the emission of greenhouse gases and to restore the ozone layer. This, the most exciting and demanding job I have ever had, does not leave me much spare time to campaign on fluoridation. However, I have managed to write this chapter in my holidays.

CONCLUSION AND LESSONS

As a scientist who tries to work for the community, I have over the years had to confront several powerful industries and interests. In my view the fluoridation establishment has been more influential and more misleading in the information it provides than the uranium/nuclear power industry.

In challenging the establishment 'experts' on fluoridation and other issues, I have found that both grassroots opposition and anti-establishment 'experts' are necessary. Without the former there is no community base and no political pressure for stopping fluoridation, and without the latter the movement would have much less credibility with the media, other professionals or scientists and decision makers.

The profluoridation establishment is aware of the danger to their power and influence from anti-establishment 'experts.' My own experience, and that of other anti-fluoridation scientists, medical doctors, and dentists, has exposed the following techniques used by the establishment for suppressing scientific and public questioning of fluoridation and for damaging the credibility of anti-establishment experts:

- the production of misleading information (e.g., see Table 3.1) for distribution to decision makers and the public;
- *de facto* censorship of scientific, medical and dental journals, by pressuring editors to send manuscripts which raise awkward questions about fluoridation to hostile referees who are establishment 'experts';
- intimidating into silence dentists, medical doctors and scientists who have concerns about fluoridation, by means of:
  - personal attacks, and misrepresentation of the fluoridation critics' work, in the media and professional journals;
  - damage to the career prospects of critics through professional associations and employers;
- keeping informed opposition out of the press/media by informing journalists and editors that:
  - opponents are either cranks, right wing extremists or alternative health 'faddists';
  - the issues being raised have already been considered twenty years ago and are therefore not news;
  - publishing or broadcasting anything on the issue would be damaging to public health;
  - fluoridation is endorsed by the WHO, USPHS, NH&MRC, AMA, ADA, and the like.

All except the last of these claims are false. In the latter claim, it is mainly small elites within the listed organisations which have actually endorsed fluoridation.

- if critics of fluoridation somehow manage to get media coverage, ensuring that a pro-fluoridation 'expert' always has the right of reply and if possible the final say; and then publicly attacking the motivations and qualifications of critics;
- circulating covertly, to decision makers and media, dossiers and reports attacking opponents personally or by association and misrepresenting their work on fluoridation.

Until recently, these tactics by the pro-fluoridation establishment successfully stereotyped the opposition to fluoridation and intimidated some opponents, thereby creating a barrier of silence in the dental and medical literature and in the popular media. An outcome is that two-thirds of Australians and half of New Zealanders, US Americans and Irish drink fluoridated drinking water. These human guinea pigs are at risk of developing skeletal fluorosis, hip fractures, hypersensitivity or intolerance reactions, and dental fluorosis. It may also turn out that they risk damage to the immune sys-

TABLE 3.1. Some Mystifications by Fluoridation Proponents

Mystification or Propaganda	My Response
Fluoride is a natural substance and so it must be safe.	Some natural substances are harmful, even those found naturally in drinking water (e.g., radium). There is scientific evidence that both radium in above-average concentrations in drinking water and natural fluoride at 1 ppm in drinking water are harmful.
Fluoride is a natural substance and so is not a medication.	Many medications are or were originally natural substances: for example, penicillin, digitalis, salicilates (in aspirin). Since fluoride is used to treat people rather than to purify the water, it is a medication and so should not be taken unless the dose is controlled.
Fluoride is an essential nutrient and tooth decay is caused by a "deficiency of fluoride."	Fluoride in doses of 1 mg/day is neither necessary for life nor for sound teeth. Even at much lower doses, nobody has ever been able to show that there is a nutritional requirement for fluoride. Any small benefit of fluoride in reducing tooth decay arises from its action on the surface of teeth.
Fluoride strengthens bones and so is a valuable treatment for osteoporosis.	Fluoride increases bone mass in a disordered way, making bones more brittle. There are now several major epidemiological studies from the United States and Britain showing a higher rate of hip fracture in the aged living in fluoridated areas than in unfluoridated areas. Moreover, treatment of osteoporosis with high doses of fluoride has been discontinued in most places.
The fluoride concentration in drinking water is controlled to within plus or minus 20 percent.	It is the fluoride dose (e.g., in mg/day), not the concentration in mg/liter, which determines the health hazards. The dose depends on the amount of water drunk and so cannot be controlled.
The bone/joint disease skeletal fluorosis is only seen in areas where drinking water contains more than 8 ppm fluoride.	In India, skeletal fluorosis is quite common when the (natural) fluoride concentration in drinking water is less than 2 ppm, and has even been reported in a few locations where it is as low as 0.7 ppm.
To suffer ill-effects from fluoride, one would have to drink a bathtub full of fluoridated water.	This confuses acute toxicity from a single high dose of fluoride with chronic toxicity from many low doses. Over a lifetime spent in a fluoridated area, one consumes and stores in the bones much more fluoride than that contained in a bathtub full of fluoridated water.

tem, genetic damage, and bone cancer, but the latter three issues have not as yet been resolved.

Since the late-1970s, the tide has slowly begun to turn. First, the implementation of fluoridation of community water supplies has almost ground to a halt as a consequence of the efforts of the community based anti-fluoridation movement, assisted by a few non-establishment 'experts.' The curtain of silence has been torn in many places, most notably in Australia and New Zealand. This has been mainly the result of efforts the determination of a few dentists, medical doctors, scientists, and other scholars scattered around the world. I think that my own greatest impact on opening up the fluoridation debate has been through the publication of my paper in the leading international science journal, *Nature*, and the associated media publicity it gained.

Further progress in rolling back fluoridation will come from building alliances with the consumer, environmental and community health movements, and by continuing to present the evidence of concern to uncommitted scientists and health professionals. The original power of the pro-fluoridation establishment, its foundation of hierarchical endorsement, is also its greatest weakness. As the silence barrier is broken in more places, more health professionals and dentists will become better informed about the issue and more of these will dare to voice publicly their doubts about fluoridation.

I hope that this exposé of the fluoridation establishment and its tactics will assist in that process. However distasteful it may seem, the public exposure of intellectual suppression is the best way of countering it.[59] As the suppression is illuminated and destroyed, the fluoridation of drinking water will come to be recognised as the harmful aberration that it is.

APPENDIX. Outline of My Critique of Fluoridation

As I see it, the case against fluoridation has three dimensions: scientific (risks and alleged benefits), political (including the establishment power structure and sources of funding) and ethical.

At the beginning of 1990, my scientific position on the alleged benefits and health hazards of water fluoridation was given in some detail in two major review papers.[60] Before then, a valuable review was published in *Chemical & Engineering News*[61] and still earlier the detailed classic book by Waldbott, Burgstahler, and McKinney.[62] Since 1990, important new scientific evidence has been published on the role of fluoride in increasing hip fractures in older people and possibly bone cancer in rats (see below). On the politics and sociology of fluoridation, I recommend the books by Varney[63] and

Martin[64] respectively; a brief account is also given in the paper by Diesendorf and Varney.[65] On the ethics of fluoridation, I wrote a paper in 1989 which I am still trying to publish in a 'respectable' journal.

Established Health Hazards

Dental fluorosis, skeletal fluorosis, hip fractures, and hypersensitivity/intolerance reactions (see text).

Note (1): Most of the major cities of Australia were only fluoridated in the 1960s and 70s, and so by 1992, older Australians had only ingested fluoridated water typically for fifteen to twenty-eight years. Both skeletal fluorosis and hip fractures will be much more prevalent in artificially fluoridated areas in the future when people have been exposed to fluoridated drinking water from birth to old age.[66]

Note (2): The prevalence and severity of dental fluorosis are increasing in fluoridated countries where they have been monitored (i.e., United States and New Zealand).[67]

Possible Health Hazards

In addition to the above established health hazards, which are each confirmed by several independent studies in the medical or scientific literature, there is evidence that the following may also be health hazards, but this has not yet been proven beyond reasonable doubt.

Cancer: In 1990, a study by the U.S. National Toxicology Program found that a small fraction of laboratory rats which ate fluoride developed bone cancers, but not any in the control group which ate much lower amounts of fluoride.[68] The results of this study were officially labelled as 'equivocal' (although this is contested by independent scientists) and other studies are in progress. Most epidemiological studies of human populations have not been able to establish a link between fluoride and cancer when differences in age, sex, and race are included properly, but an important study by Erickson is an exception.[69]

Damage to the immune system.[70]

Hazards to formula-fed babies: There is a natural physiological mechanism which stops almost all fluoride ingested by mothers from entering breast milk. The result is that babies which drink milk formula made up with fluoridated water consume over one hundred times the fluoride ingested by breast fed babies. So, people who were fluoridated as babies are likely to be at higher risk of developing the above diseases.[71]

Exaggerated Benefits

Until quite recently, it was claimed by proponents that fluoridation reduces tooth decay in children by 50 to 70 percent compared with that in unfluoridated areas. In general, the studies which were supposed to support this large alleged reduction tended to be conducted by enthusiasts for fluoridation and their scientific quality was very low. Not one was a time dependent study with randomly chosen test and control populations and 'blind' examination of teeth.[72] The reports of some studies claiming large benefits from fluoridation were so misleading that questions of possible fraud have been raised.[73]

Another means of overestimating benefits came from pro-fluoridation studies which compared large, fluoridated cities with small unfluoridated rural towns. This is an inappropriate comparison, because diet is often worse and tooth decay higher in rural areas. But, by comparing major cities we can reduce dietary differences. Then we find that tooth decay in Australia's only unfluoridated major city, Brisbane, is about the same as in fluoridated Adelaide and Perth, and is less than in fluoridated Melbourne.[74] In New Zealand, tooth decay in unfluoridated Christchurch is about the same as that in all the other major cities of that country, which are fluoridated.[75] Similar results have been reported from the United States, Canada, and elsewhere.[76] Nowadays there is little or no significant difference in tooth decay in permanent teeth between many *comparable* fluoridated and unfluoridated regions.

Furthermore, the pro-fluoridationists' attempt to explain the low tooth decay in unfluoridated Brisbane and Christchurch as resulting from imported soft drinks processed in fluoridated areas, is unconvincing, because Brisbane and Christchurch are so large and isolated that these cities manufacture most of their own soft drinks, or just import the concentrate but not the water.

Recently, some proponents have admitted that the benefits of fluoridation are now considerably less than the alleged 50 to 70 percent reductions in tooth decay, for example, only 20 percent reduction. Nowadays, in an average ten year old Australian, this corresponds to only one-fifth of a dental cavity, which is negligible.

There are well-designed experiments[77] which show, beyond reasonable doubt, that fluoride toothpaste is effective in reducing tooth decay. But, fluoride toothpaste has about one thousand times the fluoride concentration of fluoridated water, so we cannot deduce from its effectiveness that fluoridated water is also effective. There is now a large body of evidence that fluoride at sufficiently high concentration acts on the surface of teeth to reduce tooth decay, but there is little or no benefit from actually ingesting fluoride (see Box 2).

In most western countries, tooth decay has declined substantially in *unfluoridated* regions over the past two to three decades. In several cases—such as Sydney, Australia; New Zealand; Gloucestershire, United Kingdom; and parts of Canada—this decline commenced at least several years before water fluoridation was introduced. But fluoridation was often wrongly given the credit.[78] Other factors which could be responsible for the declines in unfluoridated areas are dietary changes, improved dental health education and toothbrushing habits, fluoride toothpaste (in the 1970s, but not before) and changes in immunity.

In support of dietary changes as an important factor, there is now scientific evidence that chewing cheese reduces tooth decay. In Australia, the consumption of cheese increased substantially from the 1950s to the 1980s, spanning the period of declining tooth decay.

Politics

Fluoridation has been heavily funded by the aluminium and sugary food industries, which have vested interests in the image of fluoride as a safe and effective reducer of tooth decay.

Aluminium smelters benefitted both directly and indirectly from fluoridation. Initially they sold their fluoride wastes to water authorities[79] and, once the image of fluoride was changed from that of a pollutant to a beneficial dental/public health chemical, they obtained decades of relief from pollution controls. The latter was the principal payoff for that industry.

The sugary food industry gains sales from the notion that there is a magic substance in drinking water which reduces tooth decay, whatever sugary food our children may eat. In the United States, research on diet, nutrition and tooth decay has been funded by the Sugar Research Foundation, enabling the industry to exercise some control over the direction of research and the production of results which could embarrass it. In Australia, the Dental Health Education and Research Foundation, one of the main fluoridation promoting bodies in New South Wales, has been funded by Coca-Cola, Colonial Sugar Refining Co., Cadbury-Schweppes, Australian Council of Soft Drink Manufacturers, Kelloggs (sugary processed cereals), and Scanlens (sweets), among others.[80]

Academic dentists and dental public health officials gain promotion for themselves and status for their professions by promoting the fluoridation of water supplies as a public health measure. Bodies like the Australian Dental Association and the National Health and Medical Research Council have been claiming since the early 1950s that "fluoridation is safe and effective." Now they seem unable to give unbiased consideration to scientific data showing that they were wrong.[81]

Ethics

Fluoridation is mass medication with an uncontrolled dose with a chemical which is expensive to remove (see text).

NOTES

1. B. Martin, *Scientific Knowledge in Controversy: The Social Dynamics of the Fluoridation Debate*, Albany: State University of New York Press (1991), Appendix.

2. CSIRO, the Commonwealth Scientific and Industrial Research Organisation, is the Australian Government's national research organization. For several years around 1980, I was a Principal Research Scientist and leader of the Applied Mathematics Group in CSIRO.

3. M. Diesendorf, "On being a dissident scientist," *Ockham's Razor 2*, Sydney: Australian Broadcasting Corporation (1988a): 9–14.

4. M. Diesendorf, ed., *The Magic Bullet: Social Implications and Limitations of Modern Medicine, an Environmental Approach*, Canberra: Society for Social Responsibility in Science (1976).

5. M. Diesendorf and B. Furnass, eds., *The Impact of Environment and Lifestyle on Human Health*, Canberra: Society for Social Responsibility in Science (1977).

6. M. Diesendorf, ed., *Energy and People: Social Implications of Different Energy Futures*, Canberra: Society for Social Responsibility in Science (1979).

7. M. Diesendorf, "Have the benefits of water fluoridation been over-estimated?" *International Clinical Nutrition Review*, vol. 10, no. 2 (1990a): 292–303.

8. For instance, members of the alternative health movement sometimes claim incorrectly that, while artificially fluoridated water is harmful, naturally fluoridated water is safe.

9. For a recent review, see M. Diesendorf, "The health hazards of fluoridation: a re-examination," *International Clinical Nutrition Review*, vol. 10, no. 2 (1990b): 304–321.

10. For example, see M. Diesendorf, "International symposium on fluoridation," *Social Science and Medicine*, vol. 27, no. 9 (1988b): 1003–1005.

11. For references, see Diesendorf (1990b), op. cit.

12. O. Fejerskov et al., *Dental Fluorosis: A Handbook for Health Workers*, Copenhagen: Munksgaard (1988).

13. M. P. Crisp, *Report of the Royal Commissioner into Fluoridation of Public Water Supplies*, Hobart: Government Printer (1968).

14. These were two of my particular "science and society" interests in the 1970s.

15. P. R. N. Sutton, *Fluoridation: Errors and Omissions in Experimental Trials*, Melbourne: Melbourne University Press (1960, second edition).

16. See G. Walker, *Fluoridation: Poison on Tap*, Melbourne: Glen Walker (1982).

17. Diesendorf (1976), op. cit.

18. Not because of the brief mention of fluoride.

19. The conference proceedings were published as Diesendorf and Furnass, op. cit.

20. Published in Diesendorf and Furnass, op. cit.: 265–280.

21. Royal College of Physicians, *Fluoride, Teeth and Health*, Tunbridge Wells, Kent: Pitman Medical (1976).

22. M. Diesendorf, "Is there a scientific basis for fluoridation?" *Community Health Studies*, vol. 4 (1980): 224–230.

23. J. A. Yiamouyiannis and D. Burk, "Fluoridation and cancer: age dependence of cancer mortality related to artificial fluoridation," *Fluoride*, vol. 10, no. 3 (1977): 102–123.

24. See Appendix.

25. Before my departure, I received several phone calls from strangers who pretended to be anti-fluoridation but were clearly trying to find out my movements and those of Dr. Yiamouyiannis. When I called one of these people back, the phone was picked up by her husband who inadvertently revealed that she had given me a false name.

26. See Martin, op. cit., chapter 3.

27. G. L. Waldbott, A. W. Burgstahler, and H. L. McKinney, *Fluoridation: The Great Dilemma*, Lawrence, Kansas: Coronado Press (1978).

28. As witness the debate in the letters columns of the *Medical Journal of Australia*.

29. G. L. Waldbott, *A Struggle with Titans*, New York: Carlton Press (1965); Waldbott et al., op. cit.

30. Waldbott, op. cit.

31. This curtain of silence is still quite thick in the United States and the United Kingdom.

32. Private communications from ABC journalists.

33. I understand that this legislation was revoked recently, thanks to the efforts of the Green Independents.

34. D. H. Leverett, "Fluorides and the changing prevalence of dental caries," *Science*, vol. 217 (2 July 1982): 26–30; First International Conference on Declining Prevalence of Dental Caries, *Journal of Dental Research*, vol. 61 (1982) (Special Issue).

35. Reviewed in Diesendorf (1990a), op. cit.

36. For example, National Health and Medical Research Council, *Report of the Working Party on Fluorides in the Control of Dental Caries*, Canberra: Australian Government Publishing Service (1985).

37. Published subsequently as W. Varney, *Fluoride in Australia: A Case to Answer*, Sydney: Hale & Iremonger (1986).

38. For example, letter from N. L. Henry, Federal President, Australian Dental Association, to Barry O. Jones, Minister for Science and Technology, dated 28 August 1985, with Minister's annotations.

39. Waldbott, op. cit.

40. H. Moolenburg, *Fluoride: The Freedom Fight*, Edinburgh: Mainstream Publishing (1987).

41. Martin, op. cit., chapter 5.

42. For details and references see Diesendorf (1990a), op. cit.

43. My own view is that fluoride at 1 ppm in drinking water has at best a very small benefit as it passes over the teeth, and that it is more effective at 1000 ppm in fluoride toothpaste.

44. M. Diesendorf, "The mystery of declining tooth decay," *Nature*, vol. 322 (10 July 1986a): 125–129.

45. See Martin, op. cit.: 76.

46. The WHO committee on fluoridation contains no opponents—see Waldbott et al., op. cit., chapter 16 and Varney, op. cit.

47. Published as M. Diesendorf, "A re-examination of Australian fluoridation trials," *Search*, vol. 17 (1986b): 256–262; M. Diesendorf, "Anglesey fluoridation trials re-examined," *Fluoride*, vol. 22, no. 2 (1989): 53–58.

48. I have given a detailed account of this exciting symposium-debate in Diesendorf (1988b), op. cit.

49. For example, the article claimed falsely that "When the data [Dr Colquhoun's] was [sic] re-analysed for previous fluoride exposure by the NZ Medical Research Council, Colquhoun's 'findings' evaporated." But, when Dr. Colquhoun referred this passage to the Director of the Medical Research Council of New Zealand, he replied: "You are right in your assumption that this Council has not at any stage set out to re-analyse your research data, nor has it contracted others to do so."

50. Anon, "Fluoridation disaster in the A.C.T.," *ADA News Bulletin*, no. 162 (November 1989): 7–8. Replies by M. Diesendorf, *ADA News Bulletin*, no. 166 (March 1990): 6, 8, and J. Colquhoun, *ADA News Bulletin*, no. 167 (April 1990): 17–18 .

51. J. Colquhoun and R. Mann, "The Hastings fluoridation experiment: science or swindle?" *The Ecologist*, vol. 16 (1986): 243–248; J. Colquhoun and R. Mann, "The Hastings fluoridation experiment:

postscript," *The Ecologist*, vol. 17 (1987): 125–126; M. Diesendorf, "Misleading publicity for a fluoridation trial," *New Zealand Medical Journal*, vol. 101 (13 December 1988c): 832–833.

52. National Health and Medical Research Council, *The Effectiveness of Water Fluoridation*, Canberra: Australian Government Publishing Service (1991).

53. Out of more than a dozen of our scholarly papers and books on fluoridation, the report cited only one and that was published in a popular journal, *The Ecologist*, and so has less scientific status.

54. But the report obscured the evidence we presented that water fluoridation, rather than fluoride toothpaste or fluoride tablets, was often the principal source of overdosing.

55. C. Cooper and S. J. Jacobsen, "Water fluoridation and hip fracture," *Journal of the American Medical Association*, vol. 266, no. 4 (1991): 513–514; S. J. Jacobsen et al., "Regional variation in the incidence of hip fracture: U.S. white women aged 65 year and older," *Journal of the American Medical Association*, vol. 264 (1990): 500–502; M. F. R. Sowers et al., "A prospective study of bone mineral content and fracture in communities with differential fluoride exposure," *American Journal of Epidemiology*, vol. 133 (1991): 649–660; C. Danielson et al., "Hip fractures and fluoridation in Utah's elderly population," *Journal of the American Medical Association*, vol. 26 (1992): 746–748; H. Jacqmin-Gadda et al., "Fluorine concentration in drinking water and fractures in the elderly," *Journal of the American Medical Association*, vol. 273 (1995): 775.

56. Colquhoun and Mann (1986), op. cit.; Diesendorf (1988c), op. cit.

57. A. J. McMichael and G. D. Slade, "An element of dental health? Fluoride and dental disease in contemporary Australia" (editorial), *Australian Journal of Public Health*, vol. 15, no. 2 (1991): 80–83. Replies by M. Diesendorf and by J. Colquhoun, *Australian Journal of Public Health*, vol. 15, no. 4 (1991): 308–310.

58. See Diesendorf (1990a), op. cit. and Diesendorf (1990b), op. cit.

59. B. Martin, C. M. A. Baker, C. Manwell, and C. Pugh, eds., *Intellectual Suppression: Australian Case Histories, Analysis and Responses*, Sydney: Angus & Robertson (1986).

60. See Diesendorf (1990a & b), op. cit.

61. B. Hileman, "Fluoridation of water," *Chemical & Engineering News*, vol. 66, no. 31 (1988): 26–42.

62. See Waldbott et al., op. cit.

63. Varney, op. cit.

64 . Martin, op. cit.

65. M. Diesendorf and W. Varney, "Fluoridation: politics and strategies," *Social Alternatives*, vol. 5, no. 2 (1986): 48–53.

66. See Diesendorf (1990b), op. cit.

67. J. Colquhoun, "Disfiguring dental fluorosis in Auckland," *Fluoride*, vol. 17 (1984): 234–242; Diesendorf (1990a), op. cit.

68. National Toxicology Program, *Toxicology and Carcinogenesis Studies of Sodium Fluoride in F344/N Rats and B6C3F1 Mice*, Bethesda, MD: National Institutes of Health (August 1990).

69. When I was a member of the CSIRO Division of Mathematics and Statistics, I worked through the evidence on fluoridation with my statistician colleagues and verified that the controversial study by Yiamouyiannis and Burk, op. cit., which claimed that there is a link between fluoridation and cancer in human populations, did not adjust adequately for the different age, sex, and race distributions in the fluoridated and unfluoridated cities. But an epidemiological study by a pro-fluoridationist, Erickson, did show a clear correlation between fluoridation and cancer when age, race, and sex had been allowed for properly [J. D. Erickson, "Mortality in selected cities with fluoridated and non-fluoridated water supplies," *New England Journal of Medicine*, vol. 298 (1978): 1112–1116]. In his paper, Erickson made further adjustments for population density and median education, and these nonstandard adjustments removed the original correlation between fluoridation and cancer. Strangely, the author stated in the conclusion of his paper that "There was no evidence of a harmful effect, including cancer, attributable to fluoridation."

70. J. Yiamouyiannis, *Fluoride: The Ageing Factor*, Delaware, Ohio: Health Action Press (1986); S. L. M. Gibson, "Effects of fluoride on immune system function," *Complementary Medical Research*, vol. 6, no. 3 (1992): 111–113.

71. Diesendorf (1990b), op. cit.

72. Diesendorf (1986b), op. cit.; Diesendorf (1989), op. cit.

73. See Colquhoun and Mann (1986) and Diesendorf (1988c), op. cit.

74. Diesendorf (1990a), op. cit.

75. J. Colquhoun, "Child dental health differences in New Zealand," *Community Health Studies*, vol. 11 (1987): 85–90; J. Colquhoun, "Is there a dental benefit from water fluoride?" *Fluoride*, vol. 27 (1994): 13–22.

76. Reviewed in Diesendorf (1990a), op. cit.

77. That is, randomized, double-blind, controlled trials.

78. Diesendorf (1986) and Diesendorf (1990a), op. cit., and J. Colquhoun, "Decline in primary tooth decay in New Zealand," *Community Health Studies*, vol. 12 (1988): 187–191.

79. Nowadays fluoride used in water supplies is mostly obtained as a waste of the fertiliser industry.

80. Diesendorf and Varney, op. cit., and Varney, op. cit.

81. Diesendorf and Varney, op. cit., and Varney, op. cit.

EDWARD S. HERMAN

4

Terrorism: The Struggle against Closure

INTRODUCTION

In one important respect confronting the experts on a subject like terrorism is more difficult than on issues like fluoridation or nuclear power. On the latter topics, the public's health and safety are clearly and directly at stake, its interest in rational inquiry is evident, and anti- or nonestablishment experts or spokespersons, while at a serious disadvantage in reaching the public, can sometimes be heard widely and exert influence.[1] In the case of terrorism, where mainly distant and hazy foreign enemies are claimed to be posing a threat, the public's interest is more remote, its knowledge is slight, and it is therefore more easily caught up in and manipulated by a web of symbols. For example, political leaders in the United States, with the help of the mass media, have easily mobilized a consensus on the dire threat posed by a demonized foreign enemy like Libyan leader Muammar Kadaffi,[2] that has given them political and popular backing for attacks on Libya and indirect support for larger political agendas.[3]

This consensus has been quickly established, and alternative definitions and ways of looking at terrorism have been extremely difficult to introduce into discussions of the subject. This process of closure occurs not only because of the symbolic power of the demonized enemy, but also because the mainstream media confine themselves to an exceptional degree to official sources and establishment experts. Given the rapid consensus, unaccredited experts would hardly be understood, would elicit protests by vocal groups (including the government itself), and their participation in public forums is often vetoed in advance by officials and establishment experts, who don't like open debate any more than Commissars (see below, under "The Marginalization Process").

It is not at all difficult to deconstruct and reveal terrible flaws and bias in the writings of the establishment terrorism experts; their work is often extremely crude, rhetorical, and bears little resemblance to serious social science scholarship, so that refuting them generally involves merely looking at obvious sources and using the rules of logic, as I describe below. But their work, though technically vulnerable, is immune to critical attack by virtue of the closure process and exclusion of dissenting views. Neither my occasional collaborator, Professor Noam Chomsky, nor I have ever had an opinion column or article in the *New York Times.* I had a single opinion column on terrorism in my home town newspaper, *The Philadelphia Inquirer,* in 1983, after which I was blacklisted for the next decade. In the United States, dissident experts on terrorism have been restricted almost entirely to reaching audiences of fifty to two thousand in public gatherings, and by writing articles in small circulation journals and books that reach thousands, but in the aggregate with direct access to substantially less than 5 percent of the public.[4]

It is a cliché of the West that under free institutions, truth will eventually conquer falsehood and correct error; but in the terrorism field the question must be asked: what if unconventional views are systematically marginalized by the free institutions throughout the periods when they are socially relevant?

TERRORISM AND ITS POLITICIZATION

The *Webster's Collegiate Dictionary*'s definition of terrorism captures both the vagueness and historical scope of usage of the term: "a mode of governing, or of opposing government, by intimidation." "Mode of governing" by intimidation is "state terrorism," and for a long time the word conjured up images of the mobs and guillotines of the French Revolution's "terror." In the twentieth century, the use of extreme violence by Hitler's Nazi Germany, Mussolini's Italy, and Stalin's Soviet Union reinforced the tie-in of state and terrorism.

An alternative vision of terrorism emerged in the late nine-teenth century, which pointed to alienated and radical individuals and small groups who used violence to disrupt the established order. Here the image was the bewhiskered, fanatical looking, foreigner (earlier, Jewish or eastern European, more recently Middle Eastern), although the phrases "state terrorism" and "terrorist state" have remained in use. Thus, in a speech given on July 8, 1985, United States president Ronald Reagan denounced state terrorism and listed the cast of villains on the world stage as Iran, Cuba, North Korea, Libya, and Nicaragua.

Reagan's list illustrates the enormous politicization in the use of the word terrorism. The named villains were all states with which the United States was in conflict. Nicaragua was actually under attack by a U.S.-organized and funded proxy army (the *contras*), and was therefore a literal victim of U.S.-sponsored terrorism,[5] but its designation as one of the world's terrorist states was presented in the mainstream media without comment in news stories and editori-als. South Africa, which was supporting its own cross-border insur-gents in Angola (Savimbi and UNITA) and Mozambique (RENAMO), and engaged in regular commando raids and invasions across the borders of the front line states, was not designated a terrorist state, nor was Israel, which had invaded Lebanon in 1982, maintained a cross-border proxy army in Southern Lebanon, and carried out fre-quent air and commando attacks on Lebanon.

As a further reflection of the politicization of usage, the Soviet Union, Syria, and Iraq were excluded from Reagan's 1985 list, although the Soviet Union's alleged centrality in world terrorism was repeatedly cited by the U.S. Secretaries of State and Reagan himself, and was a core element in the western ideology of terror-ism elaborated in the 1980s. But the U.S. president was about to meet with the Soviet head of state, so the Soviet Union was momentarily exempted from status as a terrorist state for diplo-matic reasons. Syria had just helped the United States win the release of hostages in Lebanon, so it too was relieved of terrorist state onus as a reward for services rendered. Saddam Hussein's Iraq was also exempt from terrorist status, despite its ongoing aggres-sion against Iran, employment of chemical weapons in the Iran war and against its own Kurdish people, and more general repres-sion at home. But Iraq's aggression against Iran was approved and aided by the West, and Iraq was, like South Africa and Guatemala, "constructively engaged," not treated as an aggressor or terrorist state. It was only when it attacked the wrong victim (Kuwait) that the U.S. officials spoke of "naked aggression," and Iraq was reclassified as a terrorist state.

It should also be noted that in western semantics, countries were not classed as "terrorist states" if they merely abused their own citizens, but only if they sponsored terrorist groups outside their own borders. Thus states like Argentina, Chile, El Salvador, and Guatemala, which carried out indigenous holocausts in the 1970s and early 1980s, were not terrorists but merely indulging in "human rights" abuses, in the memorable distinction made by Secretary of State Alexander Haig in January 1981. Haig went on to disclose that the United States was going to shift its attention from "human rights" to "terrorism," claiming that the latter was a more serious problem and even an extreme version of human rights abuse. But this was a gross misrepresentation of fact. Nobody but Haig has ever claimed that terrorists, in the narrow sense in which he used the term, have intimidated and killed on the scale of state terrorists. The 13-year total, 1968–1980, for world-wide terrorist killings, given by the CIA in 1981, was 3,680, a figure exceeded by Guatemalan government killings by more than twenty-fold between 1978 and 1983.[6]

The fact is that the Reagan administration was actively supporting state terrorists in Argentina, Chile, El Salvador, Guatemala, and South Africa (among others) in the early 1980s. The Reaganite redefinition of terrorism so as to exclude the state terrorism of its clients was thus an arbitrary politicization of the word, a manipulation of language to serve an immediate political agenda. The Reaganites did want to capture *some* states in the terrorism web, however, so they retained the notion of state terrorism in the form of possible sponsorship of "international terrorism" across borders. The world's terrorists therefore included the various governments which aided individuals, groups, and insurgencies labeled terrorist, the supporting governments being "sponsors" of terrorism. This convenient lexicon permits the invidious word "terrorist" to be applied to anybody using force against the West, or helping those who do so. The latter were part of *The Terror Network*, as set forth in a 1981 book by Claire Sterling, who tied all the left and insurgent groups of the world to a Soviet support system.

A further problem for the new lexicon was how to exclude numerous groups like the Cuban refugee network in the United States, UNITA in Angola, supported by South Africa and the United States, the *contras* attacking "soft targets" in Nicaragua from Honduras, under U.S. sponsorship, that fit the mainstream notion of terrorists perfectly. The Reagan administration handled this easily: those apparent terrorists supported by the United States and South Africa were "freedom fighters," so that neither the United States nor South Africa were terrorist states. In fact, however, the African National Congress's (ANC's) insurgency had a majority-based constituency in South Africa, whereas UNITA and the contras were essentially creatures of their foreign organizers and sponsors.[7]

THE ACCOMMODATING EXPERTS

What is most interesting is the fact that the U.S. mainstream experts and media accepted without challenge the Reagan administration's definition of terrorism and the classification of terrorists and terrorist states, despite the obvious arbitrariness and political basis of their selections. They also accepted the implicit model of terrorism in which the Soviet Union and its leading proxy, Libya, were encouraging and sustaining terrorism in order to destabilize the western "democracies" (presumably including Guatemala and South Africa). Even the truly laughable politicization of 1985 noted earlier, where Reagan listed the U.S. enemies of the moment, including the victimized Nicaragua, and "temporarily" excluded the Evil Empire, Syria, and Iraq for good behavior, did not evoke any comment. The terrorists were what a very opportunist state apparatus declared to be terrorists, however absurd and vacillating the designations.

Although a clear *prima facie* case can be made that the 1980s insurgents in South Africa, Guatemala, and El Salvador were victims of state terrorism and that the word terrorist should have been applied to the governments of those countries, this was not done by accredited experts in the United States and its allied countries. Thus, a study a colleague and I carried out of the work of thirty-two of the most prominent western experts on terrorism showed that thirty-one focused almost exclusively on insurgent terrorists—minus the Reagan-designated "freedom fighters," of course—along with the Reagan-approved list of state terrorists.[8] As another dramatic illustration of the bias of the establishment experts, we tabulated the index references to rightwing and leftwing terrorists in four major and reasonably representative establishment studies of terrorism,[9] including popular works by Sterling and by Dobson and Payne, and "scholarly" studies by Laqueur and Wilkinson.[10] We included in the listings both small-scale terrorists of the right and left (e.g., the

Italian Stefano Delle Chiaie and Salvadoran Roberto D'Aubuisson on the right, Arab Abu Nidal and the U.S. Weathermen on the left) and state terrorists allied with the West and those deemed enemies of the West (South Africa's Botha, Argentina's Videla versus Kadaffi and Castro). The aggregated totals of index references to non-Western and leftwing terrorists versus Western and rightwing terrorists was 733 to 2! In short, the establishment experts focused unremittingly on those that fit the United States and Western political agenda and simply ignored those who did not fit.

It should also be pointed out that the Soviet conspiracy model of terrorism, according to which the world's terrorism was a result of a Soviet destabilization plan and its implementation, was accepted by a solid majority of the thirty-two experts.[11] The "scholarly" Wilkinson, for example, castigated the Central Intelligence Agency (CIA) professionals who opposed the Soviet conspiracy model and he implicitly supported CIA head William Casey's efforts to make the CIA into a completely politicized instrument of state policy in its evaluation and presentation of data. In dealing with South Africa, Wilkinson not only failed to call South Africa a terrorist state, he suggested that the "troubles" were in good part a function of Soviet meddling. He even chided Kadaffi for giving aid to Nicaragua, under United States attack, and expressed the view that Britain owed the United States support for its past services to the Free World, making the facts in any particular case quite irrelevant.[12]

ACCREDITATION BY CONFLICT OF INTEREST

One of the most interesting facts about the dominant experts on terrorism is their rampant conflicts of interest. Of the thirty-two we studied, twenty-two had worked for a western government, including seven who had worked for the United States CIA. Another (but overlapping) group of fifteen were principals or employees of private security firms that served a government and business clientele. Twenty-three were or had been affiliated with nonprofit research and policy institutes (so-called think tanks), thirteen with the big four (American Enterprise Institute, Georgetown Center for Strategic and International Studies, Heritage Foundation, and the Hoover Institution). These thinktanks are closely affiliated with the government, but are largely underwritten by the corporate community. Given these relationships, the experts' identification of "terrorists" as those so regarded by their employers and funders was a foregone conclusion.

One would think that such relationships would rule out most of these experts from use by the media on grounds of conflict of interests. It works just the opposite in the West: *conflict of interest accredits the expert* because it demonstrates contacts, knowledge, and credentials. Bias is irrelevant if it is consistent with dominant mainstream opinion or reflects the opinion of very powerful people. Affiliation with a leftwing party or funding by an interest group that represented a nonestablishment viewpoint would be regarded as posing a conflict problem, and if those representatives were allowed to speak at all, their conflict would be mentioned. The U.S. public broadcasting system has barred union funding of programs as posing a conflict of interest problem, but corporate underwriting of rightwing economists and commentators is not seen as conflictual.[13] Thirty-year CIA veteran propaganda expert and former CIA station chief in Turkey, Paul Henze, was one of the leading commentators on the alleged KGB-Bulgarian plot to kill the Pope. Not only was his CIA affiliation not considered by the media to compromise his objectivity, it was not even disclosed to the public.[14]

The U.S. system thus works with great efficiency to get over and continually reaffirm the government-establishment preferred line on terrorism. The government view is explained by officials, who focus with great indignation on their preferred terrorists, sometimes offering outright lies.[15] The experts—accredited by their conflict of interest that assures their adherence to the government view—solemnly restate the government view, and mull over why the terrorists are misbehaving and what our anti-terrorism options are. What makes the system so efficient is its uncoerced character, with the free market, not a censoring government, filtering out unwanted facts and perspectives. And the free market participants usually do their private censoring with a certain innocence, as the biases they require from their experts are seen as the mere accepting of self-evident truths from which any intelligent debate must begin.

MY OWN WRITINGS ON TERRORISM

My own writings began during the Vietnam war years 1965–1975. They were driven by outrage at the events in progress and the U.S. establishment and media apologetics, which included some remarkable word usage, double standards, and rewriting of history.

The United States had entered Vietnamese affairs as a successor to the French colonial regime, whose return to power the United States had underwritten after World War II, until the French withdrawal in 1953. In 1954, the United States put into office in Vietnam a U.S.-trained and imported puppet, Ngo Dinh Diem. Although devoid of substantial indigenous support, Diem nevertheless got 99 percent of the vote in a 1955 election, and used increasingly ferocious tactics and weaponry against the peasant population. Despite these considerations, the U.S. media consistently supported the U.S. intrusion as reasonable and democratic in intent. Although the Diem regime and the United States relied on force and terror to achieve their ends from 1954–1975, the U.S. media used the word "terrorism" only in reference to the operations of local enemy insurgents. "Sideshows" such as the invasion and intensive bombing of Cambodia and the U.S.-sponsored coup and slaughter in Indonesia were also treated very antiseptically and without indignation, never as state terrorism.

One of my earlier books on terrorism, *Atrocities in Vietnam: Myths and Realities*, was published in 1971 by the church-supported Pilgrim Press. It focused on the different types of atrocity—insurgent shootings and bombings, B-52 and other bombing raids, and the use of napalm and chemical warfare, including crop destruction programs—addressing their levels, casualties inflicted, and legality. There was little difficulty in showing that U.S. direct and indirect atrocities were vastly larger, enormously more costly to the civilian population, and more consistently in violation of the laws of war, than those of the enemy. Even before this book was in print, the editor ran into opposition within the publishing house, and a struggle ensued over whether the book should be published in accord with the contract. It was issued, after a delay, in a very small format and small print edition. But it was not advertised and was not kept available for very long—the residual copies were destroyed, not remaindered (or even offered to the author for purchase).[16] The editor who had defended and pushed the book was fired within six months after it was issued.

An even more extreme case of suppression occurred in connection with a work on terrorism written with Professor Noam Chomsky, with whom I began to collaborate during the Vietnam War era. Our experience with suppression occurred in our dealings with Warner Modular Publications, a small subsidiary of Warner Brothers, that specialized in short dissident studies in various fields that could be used as course modules. Chomsky and I produced a module in 1973 entitled *Counter-Revolutionary Violence: Bloodbaths in Fact and Propaganda*. "Bloodbaths" were a favourite topic of those years, with much establishment concern and speculation over those that might occur under future Communist rule, but with comprehensive neglect of ongoing bloodbaths. Our monograph focused on these. We distinguished between Benign and Constructive Bloodbaths—those of the United States and its client states and allies—and Nefarious and Mythical Bloodbaths in Vietnam—which were bad and deserving of indignation and concern.

After twenty thousand copies of our book had been printed, it came to the attention of the top brass of the parent corporation. They were outraged, canceled the already arranged advertising, refused to sell the module, and shortly thereafter fired the officers of Warner Modular Publications and liquidated the organization. The stocks of our book were transferred to another company that never advertised it, so that the volume was effectively suppressed. We considered suing for breach of contract, but ended up obtaining full rights to republish. The story of this suppression was recorded in Ben Bagdikian's book *The Media Monopoly*,[17] and in our own later writings, but it was of no interest to the mainstream media or civil libertarians. In retrospect, one of the notable features of this suppression was the active participation of Mr. Stephen Ross, then head of Warner, and until his recent death the top officer of Time-Warner, the world's largest media conglomerate.

Six years later Chomsky and I published a greatly enlarged and updated version of the suppressed volume, entitled *The Washington Connection and Third World Fascism* as the first the first of a two-volume set on *The Political Economy of Human Rights*.[18] We included there "A Prefatory Note on the History of the Suppression of the First Edition of This Book," and the book elaborated on the themes of the earlier volume. In particular, it put great emphasis on the political-economic underpinning of U.S. relationships with states like Brazil, Chile, Indonesia, and the Philippines, the centrality of "favorable climate of investment," and the importance of state terror to serve that dominant objective. Along with extensive documentation of terror conditions in "The Pentagon-CIA Archipelago" (title of chapter 2), the book addressed and criticized the way in which usage of "terrorism" in the West had been nicely adjusted to exempt the West and its clients. *The Washington Connection*, published by South End Press, a recently formed nonprofit and critical publisher located in Boston, received a fair amount of attention in the dissident press and underground, and its aggregate sale of some forty thousand copies was huge for a dissident publication. The book was not reviewed in the leading mainstream media, however, and was given hostile treatment by *The Nation*, which had it reviewed by a *New York Times* reporter, A. J. Langguth, despite the book's severe criticism of his paper.

I followed this up with another volume published by South End Press in 1982, *The Real Terror Network: Terrorism in Fact and Propaganda*. In title and substance the work was a response to a 1981 bestseller by Claire Sterling, *The Terror Network*. Sterling constructed her terror network with the Soviet Union as center, with Libya, Cuba, and the PLO as major proxies, and with insurgent movements like the ANC and other left insurgencies as Soviet agents. It also carefully excluded South Africa, Argentina, and the Cuban refugee network in the United States, and in fact any non-left group or government favored by the West. This fitted precisely the demands of western power brokers, anxious to pin the Evil Empire label on the Soviet Union and to discredit groups challenging western interests (PLO, ANC, insurgencies from below in general) by tying them into a global conspiracy.[19] Sterling nowhere defined what she meant by terrorism, provided no quantitative information, but relied heavily on dramatized recountings of carefully selected terrorist incidents and testimony of western intelligence agencies. Her gullibility was enormous, and eventually CIA personnel disclosed that she had been an unknowing channel for the "blowing back" of CIA disinformation into the U.S. media.[20]

My book *The Real Terror Network* provided a formal critique of Sterling's methods and sources. I distinguished between "retail" and "wholesale" terrorism, the former the terror of individuals and small groups, the latter states. The point of this use of adjectives was to stress the fact that retail terrorists have limited capabilities for terrorizing, whereas states can intimidate and kill on a very large (wholesale) scale. The plague of human torture that grew in the post-World War II era and the growth of death squads and disappearances were state-directed phenomena. Sterling's exclusion of wholesale terror was not only a highly political choice, it missed the main subject. But insofar as the function of the political-"expert" focus on "terrorism" is to divert attention from the greater to the lesser terror, Sterling's choices were entirely comprehensible.

The greater part of *The Real Terror Network* was devoted to describing an alternative terror network of U.S. client states in Latin America, enumerating their terror practices, tracing the network to

U.S. training and support systems, and showing how this worked out in accord with a "development model" that stressed providing a "favorable climate of investment" for transnational corporations. In states with huge inequalities and pressures from below for directly helping the majority (which adversely affect profitability), a favorable investment climate required a strong dose of terror from leaders like Marcos, Pinochet, and the assorted other Latin American generals who led the fight for "stability."

This book had a great deal of quantitative information, and did not rely on anecdotes and stories from intelligence agencies as Sterling's book did. It was very well received in dissident and Third World circles, but could not be heard by the U.S. public through the mainstream media. The *New York Times*, for example, deeply interested in terrorism, "lost" half a dozen copies of the book and never did get around to even a notice of its existence. Sterling, meanwhile, was not only reviewed, but became a TV expert of choice, treated with virtual reverence, and was never asked challenging questions.

After Pope John Paul II was shot in Rome in May 1981, Claire Sterling soon surfaced again as an authority, claiming that this was a KGB-Bulgarian plot. Funded by the *Reader's Digest*, she published an article "The Plot to Kill the Pope," in the *Digest* issue of September 1982, then a 1984 book on the subject, *The Time of the Terrorists*, and was the top media expert on the subject from 1982 until after the conclusion of the trial against the Bulgarians in March 1986. The number two expert was Paul Henze, a longtime CIA official, who also wrote a book, *The Plot Against the Pope*, and became a very prominent expert in both print media and on television. Sterling and Henze collaborated with Marvin Kalb, of NBC-TV, in a major television special, "The Man Who Shot the Pope," shown in September 1982 and again in January 1983. Sterling, Henze and Kalb took Soviet-Bulgarian involvement as a foregone conclusion, based on the confessions of the assassin, a half-crazed Turkish rightist, Mehmet Ai Agca, made after many months in an Italian prison, along with a set of ideological presumptions and imputed motives and plots worthy of a James Bond movie.[21] The mainstream media swallowed the line with gusto and remarkable gullibility.

I followed the case closely and wrote a number of articles on it with a historian friend, Frank Brodhead. We eventually published a book, *The Rise and Fall of the Bulgarian Connection*, through Sheridan Square Press in New York City, timed to coincide with the end of the trial in March 1986. The essence of the book, with further updating, was also presented as chapter 4 in a book written in collaboration with Noam Chomsky, *Manufacturing Consent: The Political Economy of the Mass Media*, published by Pantheon Books in 1988. In all of these works, an effort was made to stress the remarkable convenience of the case for western political interests, and to show the implausibility of the accepted premises, the dubiousness of reliance on the claims of the accused Agca, given his background, the circumstances of his confession, and the absence of *any* evidence confirming his assertions about Bulgarian involvement. I tried to spell out an alternative explanation of why Agca shot the Pope and why the crime was being pinned on the Bulgarians and KGB. I analyzed the Sterling-Henze-Kalb model, showing the faultiness of its premises and its internal illogic, and pointing out some of the real world facts that it ignored (and which the U.S. media ignored, in lockstep). As in the case of Sterling's *Terror Network* and model, the mainstream media gave her version of the case huge and uncritical publicity and simply ignored my (or any other) counter arguments or claims. Contesting views didn't have to be refuted—they could never be aired, but had to circulate in the dissident media as de facto samizdats, without access to the general public.

The Rome trial which ended in March 1986, resulted in the release of the Bulgarian for lack of evidence, but even at this point the U.S. media allowed Sterling and company a convenient fall-back position: the case was alleged to be too "politically sensitive" for the court to find the Bulgarians guilty, and, furthermore, Italian law distinguishes between a finding of clear innocence and non-guilt for lack of evidence. Of course, an even stronger case can be made that "political sensitivity" (or convenience) and hostility toward the Evil Empire and Communist Party in Italy, which dictated the bringing of the case in the first place, made for juror bias toward finding communists guilty; and that the failure to find the Bulgarians completely innocent may have been to protect the powerful vested interests in Italy who had supported the case. Furthermore, Sterling and Henze had claimed back in 1984 that the "evidence" was virtually complete, yet a very extensive effort by the Italian state failed to produce a single piece of evidence confirming Agca's claims of Bulgarian involvement.

In a final touch, in the confirmation hearings on Robert Gates as head of the CIA in 1991, former CIA official Melvin Goodman testified to the politicization of the Bulgarian Plot case under CIA head William Casey, and added that the CIA professionals had considered the Sterling claims a farce, one reason being that the CIA had excellent penetration of the Bulgarian secret services. Following this testimony, the *Wall Street Journal*, which had given Sterling uncontested rights to peddle her views up to then, gave her a final word on a "plot." A letter which I wrote in response, pointing out her neglect of the Goodman admission of the CIA penetration of the Bulgarian secret services, was not published, nicely completing and illustrating the working of a closed "free market" system.

A final major work I wrote on terrorism, in collaboration with University of Pennsylvania PhD student Gerry O'Sullivan and published in 1990 by Pantheon Books, was entitled *The "Terrorism" Industry: The Experts and Institutions that Shape Our View of Terror*. Again, the intent was to show the one-sidedness and huge bias in the mainstream perceptions of terrorism. But the focus here was on the institutional roots of the bias. Thus, after background chapters on the western stake in "terrorism" and the model constructed to identify the West's oppositional forces as the "terrorists," the major part of the book describes the institutional apparatus—government, thinktanks, security firms, and experts—that expound and elaborate the chosen model. The phrase "terrorism industry" calls ironic attention to the fact that "terrorism" is modelled and pushed by experts who service a market demand for a certain view of terrorism, much like advertisers who push a certain brand of soap.[22]

As noted earlier, we stressed the close association of the experts of the industry with government and private business firms that have "security" problems, who constitute the "demanders" for the services of the industry. The longest chapter in the book, entitled "The Experts," provides a great deal of information on the linkages and opinions of the experts, including an extended account of the work of a dozen of the majors. We underscore the fact that the terrorism industry is multinational, as the governments, institutes, and experts of the United States, Great Britain, and Israel, in particular, but many others as well (including South Africa) have worked closely together and shared the same vision of terrorism. The book also describes how the U.S. mass media take the terrorism industry's experts as "independent" and properly accredited, allowing them to reinforce the official version of terrorism as *the* true one, providing a "natural" and beautiful closed system of discourse allowing Libya and the PLO to be the serious terrorists, with the Nicaraguan contras and the Cuban terror network ignored, and the governments of South Africa and Guatemala fighting against terrorism.

The "Terrorism" Industry was published by Pantheon books, a subsidiary of Random House, which is a subsidiary of Advance

Publications, the top company in the Newhouse media empire. Unfortunately, the book was about to go to press just at the point where Pantheon was in the process of being "defanged" by its parent, in the name of profitability considerations. The longstanding management of the noted progressive publisher André Schiffrin was replaced with a more compliant market-oriented one, and Pantheon has ceased to be a haven for quality critical books. Meanwhile, *The "Terrorism" Industry* was not advertised or pushed and the normal reaction of the mainstream media to such a critical work (mainly ignore; alternatively pan) was exacerbated by publisher confusion, disinterest, and neglect. The book fell stillborn from the press, receiving only modest attention even in the dissident media, and it was not permitted to enter in any way the national debate on terrorism. We believe that this was a function of its message and backing, not its quality.

METHODOLOGY

I obtained a Ph.D. in economics at the University of California, Berkeley, in 1953, with two minor fields, American history and economic history. As a trained economist and student of history, with a strong bent toward structural analysis and structural explanations of behavior and performance, my basic methodological approach has been, mundanely, the use of traditional scientific methods. I have long been a devotee of the work of the great French historian Marc Bloch, who put great stress on the comparative method,[23] which has seemed to me extremely useful in studying the treatment of terrorism. I have spent a lot of time exploring definitions and concepts, spelling out and analyzing models used in the terrorism field, and searching for relevant empirical evidence. This is unusual in the world of terrorism studies, where serious traditional academic investigation is sparse and badly tarnished by conflict of interest and ideological bias. A number of terrorism experts have been journalists, most often with strong ideological commitments. But even the "scholars" of the field, like Laqueur and Wilkinson, work at a huge distance from Thucydides' self-imposed guideline that "the accuracy of a report [be subject]...to the most severe and detailed tests possible."

Because of the high degree of politicization and one-sidedness of establishment terrorism accounts in the press and purported "scholarly" studies of terrorism, the questions deserving my close attention as a critic have been obvious. The manipulation of definitions to serve western interests called for a close analysis of the semantics and selectivity of usage of terrorism. The models of terrorism have been similarly structured to yield a proper political result, with the basic model of "Soviet control" and with specific models adapted for special occasions, as with the KGB-Bulgarian plot against the Pope. It has been easy to show that the definitions and models used in the mainstream studies are special cases that serve special interests—and often survive only because contesting facts and alternative models are kept under the rug.

Because of the extreme difficulty in getting an alternative view heard, I have often sought out dramatic and powerful illustrations of the state (wholesale) terror that the western terrorism industry refuses to recognize. I have frequently also emphasized the method of dichotomization and juxtaposed comparison to illustrate forcefully the fact and inappropriateness of selective western attention. For example, with the help of the Pentagon itself, during the Vietnam War, I was able to show that the United States was using ordnance in Vietnam that exceeded that used by the enemy by a ratio of five hundred to one.[24] It was not difficult to show the huge bias in the U.S. press in attention given to victims of terrorism in the Soviet Union and in U.S. client states.[25] It was not hard to demonstrate that on standard definitions of terrorism, South Africa was a far more serious terrorist state than Libya, and that—to understate the case—the "experts" and mainstream media did not recognize this fact.[26]

A great deal of the information needed for a proper study of terrorism is readily available in the reports of human rights groups like Amnesty International and Americas Watch, who cover a wide terrain, and from more specialized groups, frequently victims and expatriates, who put out newsletters and special reports on their home countries. There are also numerous books, journals, government documents, and news reports that provide valuable information. This material is available to mainstream experts; they ignore it because it doesn't fit their hypotheses and models. Because of the extensiveness of and the frequent need to actively seek out information, I have been prone to work with collaborators, to help share the research burden. The ones I have worked with have been excellent researchers, and the benefits of collaboration in all cases have greatly exceeded the costs.

In sum, writing as a critic on terrorism has been easy in that the establishment studies are so grossly biased and intellectually thin that their refutation has been like shooting sitting ducks. Counter-models have been easy to construct. Mobilizing the requisite evidence requires a great deal of digging, assembling data, and checking, but the evidence is there, sitting unused by the mainstream experts. Given the "self-evident" character of the establishment position, the reply has to be extensive, powerful, and error-free. Even then, the problem remains of how one can get heard in the mainstream, even at the most minimal level.

THE MARGINALIZATION PROCESS

I have myself "modeled" the process whereby my own writing is marginalized. Chapter 1 of *Manufacturing Consent*, entitled "A Propaganda Model," describes in detail a set of interacting forces that filter out unwanted thoughts: the ownership and control structure and profit orientation of the mainstream media; their dependence on advertising revenue; their tie-in with government as primary information source and licencer (for radio and television) and external protector (for global media); the threat of flak from the powerful; and ideological constraints. These forces press the media toward political conformity and protect establishment positions against attacks by critics who address fundamental rather than tactical errors.

One of the media's routes to safety is to confine "independent" opinion to the experts from within the establishment. This process has been carried to the extreme in the case of the "terrorism" issue, as it is largely a foreign policy matter, with properly demonized villains (Arafat, Kadaffi) and with the victims of western terrorism and double standards having no voice in the West (Guatemalan Indians, Brazilian peasants, South African blacks). Under these conditions, the system quickly closes: the government names the terrorists, the affiliated terrorism experts solemnly agree and discuss proper tactics, and the media asks no hard questions.

The experts with fundamentally dissenting views are simply left out of the discussion. They are not accredited by former government employment or affiliation with the proper thinktank or Council on Foreign Relations; that is, they don't have the requisite conflict of interest! In the usual flurry of propaganda following a terrorist incident, the government, experts, and media quickly accept as obvious the official version of terrorism. Thereafter, dissident experts would hardly be understood, as they consider the issues too selectively chosen and in urgent need of contextualization, whereas the media want commentary and debate only on the basis of accepted fundamentals (e.g., that Kadaffi is the issue, and that the problem is why he does nasty things and how we can stop him).

Only twice was I ever considered for appearance on television as a terrorism expert. In the first case, I was called by a representative of the Phil Donahue Show, who wanted to see if I qualified for an appearance. She asked me what I would suggest the United States do about terrorism. I said that the first thing was for it to stop directly doing it and indirectly sponsoring it. This left her at a loss for words, and when our conversation ended I knew that I was not going on the Donahue Show. (I realized later that I should have played dumb and not revealed my hand so early if I wanted to get on the show.) The second instance was in connection with the Plot against the Pope. I was again "felt out" by television representatives on ABC, but nothing came of it. Subsequently, however, a reporter contact within ABC informed me that Mrs. Claire Sterling refused to appear on television with anyone who would seriously oppose her views, and exercised de facto veto power over panel appearances.[27] I don't know whether that explained the particular case on ABC in which the inquiry was never followed up with an appearance. I do know, however, that Sterling refused to debate with me on the Plot against the Pope at the University of Pittsburgh, where the students offered her market rates or better to appear.

LESSONS AND REFLECTIONS

Although I have been denied any direct access to the mainstream media over the past dozen years, with minor exceptions, I do not consider my efforts futile. I have been a part of a dissident movement that depends on mutual support, including intellectual support, and I have received hundreds of letters from persons previously unknown to me who have told me that my writings "opened their eyes" or infused them with energy. The dissident movement has been a force helping contain the national establishment, and its energy and strength depends in part on raising questions, presenting inconvenient facts, and formulating alternative models of where we are, why, and where we should be going. Even marginalized intellectuals serve the containment process by strengthening oppositional forces, and their ideas sometimes trickle upward into mainstream discourse. In the Vietnam war era and Central American wars of the 1980s, fear of repercussions at home was an important element in the calculus of aggression. In the absence of criticism and protest, violence would have been greater.

In retrospect, I believe that I and many of my dissident allies have put too much emphasis on scholarly analyses and too little on reaching the general public. This is a result of the fact that many of us are academicians and gravitate easily to traditional academic modes of discourse. However, many of the issues are fairly complex, so that with our views so unfamiliar and jarring, we need more space than ten sentences on television or a 750-word opinion column to explain our position, which makes some reluctant to try. Furthermore, access to mainstream television and opinion columns often requires a struggle, and is sometimes foreclosed entirely. Nevertheless, I still think it has been a mistake to opt so disproportionately for the easy route of books and articles in dissident journals and papers with circulations of 2,000–30,000.

NOTES

1. See Diesendorf's and Sharma's chapters in this volume. An offsetting consideration is that the establishment experts in the science fields often have substantial technical expertise and qualifications. The experts on terrorism are commonly journalists, popular writers, former officials, and not very distinguished or well-trained political scientists.

2. On the demonization process, see J. Haiman and A. Meigs, "Khaddafy: Man and Myth," *African Events* (February 1986).

3. The U.S. Reagan administration, which entered office in January 1981, was notorious for using "terrorism" as a propaganda weapon for mobilizing the U.S. public for attacks on Libya, and indirectly as a means of helping justify its renewal of the arms race and as a cover for its regressive economic policies. See Edward S. Herman and Gerry O'Sullivan, *The "Terrorism" Industry: The Experts and Institutions that Shape Our View of Terror*, New York: Pantheon Books (1990): 22–25 and passim.

4. A State Department funded study in the mid-1980s, based on interviews with twenty-eight hundred former Soviet citizens living in the United States, estimated that 77% of blue collar workers and 96% of the middle elite in the Soviet Union listened to foreign broadcasts, and that underground publications were read by 45% of high level professionals and 14% of blue collar workers. (James R. Miller and Peter Donhowe, "The Class Society Has a Wide Gap Between Rich and Poor," *Washington Post National Weekly* [17 February 1986].) U.S. dissidents have a much smaller outreach

than did Soviet dissidents, and, of course, foreign broadcasts are not as important in the United States, nor do they present seriously alternative views.

5. These mercenary forces were referred to by the Defence Intelligence Agency in its 16 July 1982 "Weekly Intelligence Summary" as a "terrorist" army, before they were officially designated as "freedom fighters." This point was only disclosed in 1984 by the Council on Hemispheric Affairs, a public interest group specializing in critiques of U.S. policy in Latin America.

6. See Edward S. Herman, *The **Real** Terror Network: Terrorism in Fact and Propaganda*, Boston: South End Press (1982): 8 and passim.

7. This was more true of the *contras* than UNITA, which did have an indigenous tribal base, although one heavily dependent on and a tool of South Africa.

8. The thirty-two experts included the thirteen most frequently cited by other terrorism experts, as reported in Alex Schmid, *Political Terrorism*, Amsterdam: North Holland (1983), the sixteen most frequently cited in a large mass media sample, and eight others based on our own assessment of importance and influence. There were five individuals common to the Schmid and media sample list, giving twenty-four net, plus the eight ad hoc. See further Herman and O'Sullivan, op. cit.: 143–146, 183–190.

9. Rightwing and leftwing are imprecise political categories, "rightwing" implying conservative or reactionary and supportive of regimes of private property along with oligarchic and authoritarian rule; "leftwing" implies reformist or radical support of more equalitarian ownership and control, to be obtained by democratic or sometimes authoritarian rule and methods. In Third World disputes, the great powers of the West have often supported rightwing movements, only rarely those on the left.

10. I included Claire Sterling, *The Terror Network*, New York: Holt Rinehart and Winston/ Reader's Digest (1981); Charles Dobson and Robert Payne, *The Terrorists*, New York: Facts on File (1982); Walter Laqueur, *The Age of Terrorism*, Boston: Little Brown (1987); and Paul Wilkinson, *Terrorism and the Liberal State*, New York: New York University Press (1986). See Herman and O'Sullivan, op. cit., Table 7–4: 189.

11. Ibid., Table 7–2: 184

12. Ibid.: 176–184.

13. Pat Aufderheide, "'This program was not made possible . . .': If PBS let GM sponsor Milton Friedman, why can't unions sponsor a labor history series?," *In These Times* (5–18 March 1980): 13.

14. This was a result in good part of the fact that Henze fixed his own identification, and never mentioned CIA. Henze also regularly insisted on clearing guests with whom he would appear and questions he would be asked on television. See Edward S. Herman and Frank Brodhead, *The Rise and Fall of the Bulgarian Connection*, New York: Sheridan Square Publications (1986): 123–124 [footnote 1] and 146–159.

15. For a compendium of official lies by U.S. officials on the Sandinista government of Nicaragua in the 1980s, see Peter Kornbluh, *Nicaragua: The Price of Intervention*, Washington, D.C.: Institute for Policy Studies (1987): chapter 4. For broader compendia of official U.S. lies, see Noam Chomsky, *The Culture of Terrorism*, Boston: South End Press (1988) and *Necessary Illusions: Thought Control in Democratic Societies*, Boston: South End Press (1989).

16. Edward S. Herman, *Atrocities in Vietnam: Myths and Realities*, Philadelphia: Pilgrim Press (1971). The book is now a collector's item, for which I get periodic frantic appeals from firms specializing in finding scarce volumes.

17. Ben Bagdikian, *The Media Monopoly*, Boston: Beacon Press (1983, 1987, 1990, and 1992).

18. The second volume, also published by South End Press in 1979, was entitled *Beyond the Cataclysm: Postwar Indochina and the Reconstruction of Imperial Ideology* .

19. This was especially true of the Reagan administration in the United States and the Begin government in Israel, the latter eager to make the PLO into an agent of World Communism, thus justifying its refusal to negotiate with Palestinians. The first conference of the Israel-sponsored Jonathon Institute, held in Jerusalem in July 1979, was clearly designed to mobilize such an ideological offensive. Both George Bush and Claire Sterling attended that conference. See Herman and O'Sullivan, op. cit.: 22–25, 29–36, 104–106.

20. Gregory Treverton, *Covert Action*, New York: Basic Books (1987): 165.

21. For example, the television program argues that as Bulgaria was a police state, if Agca stopped for a period in Sofia, the secret police must have known he was there and therefore been using him. During the trial in Rome a high official of Agca's rightwing group the Gray Wolves testified that the Gray Wolves like to go through Bulgaria because with the large flow of Turkish immigrants it was easy to hide. This statement, which completely contradicted the extremely silly Kalb view, was never picked up in the U.S. mainstream media.

22. This analogy was made by the head of the Heritage Foundation explaining that rightwing institutions like his own must operate like Proctor and Gamble to get over proper ideas. See Edward S. Herman and Noam Chomsky, *Manufacturing Consent: The Political Economy of the Mass Media*, New York: Pantheon (1988): 23–24.

23. Especially, Marc Bloch, *The Historian's Craft*, Manchester: Manchester University Press (1954) and "A Contribution Toward a Comparative History of European Societies," in *Life and Work in Medieval Europe*, London: Routlege and Kegan Paul (1967).

24. See Herman (1971), op. cit., Table 2: 57.

25. See Herman (1982), op. cit., Table 4–1, "Press Coverage of Abused Persons in the Soviet Union and in Eight U.S. Client States,": 196.

26. See Herman and O'Sullivan, op. cit.: chapters 2, 7, and 8.

27. The same holds for Paul Henze: see note 14 above.

9

5

"What Price Intellectual Honesty?"
Asks a Neurobiologist

CAREER

I was born in London, although both of my parents went to Scottish Universities. I was brought up to respect academics, and to believe that their paramount interest in life was the pursuit of truth. My intention had always been to take up a career in a university, in which—as long as one carried out the teaching duties assigned by the head of department—one was free to do research in any area in one's discipline, which seemed exciting. At that time in Britain academics were protected from those with orthodox opinions in power by long established tenure.

I obtained a scholarship to University College School, London, and took a medical degree at Middlesex Hospital Medical School in 1956, since when I have practiced as a part-time physician. I proceeded to a degree in neurophysiology and biophysics at the Department of Physiology, University College, London, and completed it in 1958. I also obtained a diploma in biophysics from Kings College, London.

I then went to the Institute of Psychiatry, first as a research assistant and then as an Honorary Lecturer in Biochemistry, and I stayed there until 1962. I was working on the electrical properties of slices of brain, and was invited to examine similar properties of nerve cells dissected out by hand, in Sweden. I returned to Britain in

1964 and took up the position of Biochemist and Honorary Lecturer in Applied Neurobiology at the Institute of Neurology in London. The following year I was appointed Senior Lecturer in Physiology at Battersea College and, in 1968, I was made a personal Reader in the University of Surrey. I was in charge of all physiology teaching in the University at that time, and have been the senior physiologist since then. In 1970, I set up the Unity Laboratory of Applied Neurobiology, which I have directed ever since. I have published about 150 full-length publications in cytology, neurobiology and resuscitation, and have written five books.

Throughout my career, my upbringing and training led me to entertain the following assumptions: academics' first priority is to seek the truth as they define it; they are prepared to enter into dialogue about their beliefs and research; they believe that evidence and reasoning should take precedence over belief and emotion; they behave fairly in argument; they do not practice casuistry; and they do not use power to defend their views. I have reluctantly come to the conclusion that these assumptions are not always warranted.

A STUDENT OF THE INSTITUTE OF PSYCHIATRY, 1958–1962

My first job in 1958 was as research assistant to Professor Henry McIlwain, at the Institute of Psychiatry, London. He was the most active exponent of the use of thin slivers cut from brain for the study of the biochemistry and physiology of the intact living brain. I believe that he had learnt the technique from Professor Sir Hans Krebs, the Nobel Laureate in biochemistry, with whom he had worked in Sheffield, England. The properties of the brains of adult animals could be studied in slices for up to two hours before they degenerated. Professor McIlwain had built up the Department of Biochemistry with a small nucleus of permanent staff, and ten to fifteen visiting research workers and students for doctorates of philosophy.

At the Institute of Psychiatry, I became aware of certain fairly common practices. Some people did not quote authors they did not like personally, or others who had predated them, or had findings that senior staff did not like. They would fail to do control experiments or would discard results which gave different results from those they expected. When I first heard about these practices, I was shocked by them, but I was even more shocked by the tolerance and cynicism which some of my colleagues displayed towards them.

In 1958, I was discussing a particular biochemical problem with a senior Hungarian biochemist, who came to the Institute legally before the Revolution and had applied for asylum in Britain. This was granted on condition that he remained in the same position. He told me that he quite agreed with me, but would not say so in public, as it might risk his appointment here.

I devised a simple technique for cutting slices quickly, so that they could be studied sooner than by the previous technique. One of my colleagues liked the idea, and was put on to the task of studying the properties of slices cut in this way. We worked harmoniously together, until one day he stopped coming to my laboratory, and I noticed that he was avoiding me in the corridor. A senior colleague had ordered my friend to work with him rather than to develop my technique, but I was too junior to be able to protest.

I heard along the grapevine that this senior colleague was writing up a paper based on my work, and when I asked if I would be a co-author, I was told that my help would be acknowledged. In the event when the paper appeared, I noticed that it ended with an acknowledgement for my "help" with the technique, but I had not been asked to be a co-author.

When I read the published paper, I noticed an important mistake in one of the calculations. I pointed this out politely, but I was told that it was just a different way of expressing the value. No, I maintained that this was not true, because a similar constituent had been calculated correctly, and the result reported was in impossible units.

This mistake, once published, became the correct values, and later publications showed similar ones. In a book written later, different pages show different values for the same parameter. The book is very authoritative. I did not wish to hurt the feelings of those responsible, but when I wrote a paper subsequently on the same subject, I inserted the correct calculation; there was some difficulty in publishing it.

I learned several lessons from my time at the Institute of Psychiatry. Firstly, doctoral students have no redress against their supervisors, since their careers would be ruined if they made determined criticisms, or resigned their studentships. Furthermore, well-known academics find it relatively easy to publish in a journal, especially if they are on the editorial board. Thirdly, the data reported in weighty books acquire an authority and inertia, which encourages some other people to find similar results, and dissuades others from submitting different results for publication. This was my first personal experience of misdemeanor, and it disturbed me greatly.

RESEARCH FELLOW IN GÖTEBORG, SWEDEN, 1962–1964

In 1962, Professor Holger Hydén, at the Institute of Neuro-biology, Göteborg, invited me to do similar studies on single nerve cells dissected out by hand from the brains of freshly killed rabbits.[1] The technique is brilliant in its simplicity, and is not difficult.[2] A large number of experiments were carried out on the biochemistry and anatomy of the cell bodies, but the same question was asked, as had been asked about cerebral slices. How many of the electrical properties of the living nerve cell survived its separation? I was employed with many others to find out.

I had a very interesting and fruitful time at the Institute of Neurobiology under Professor Hydén, who was very kind to me personally. However, I saw there two practices of which I had been previously unaware. A technician would produce a table of results, and the supervisor would strike out some of them, without giving any reason for so doing; the technician then retyped the table, discarded the original, and the new table became the raw data. In recent times in Britain, I have seen research workers simply deleting values from the computers attached to their instruments. I do not believe these practices are widespread in Sweden or in Britain.

The real worry was that such selected and, therefore, misleading data should occur in published papers, and then become part of the canon of knowledge. In the real world, the idea that they would be corrected when other people tried to reproduce the experiment is just wishful thinking. As a consequence of these observations, I made the decision that I would never myself indulge in, or put my name on publications, in which I knew that manipulation of results or intellectual casuistry had occurred. During the next few years, I identified a large number of widely practiced and tolerated misdemeanors, such as not discussing results which disagreed with one's own, avoiding doing crucial control experiments, answering important questions with artful circumlocutions, and so forth.[3]

Adenosine triphosphate (ATP) is one of the most important chemicals in the body. It is required for many metabolic reactions and it helps to synthesize proteins. Its energy is used to move water and it causes muscle to contract.

An American research worker visiting Göteborg, Dr. Joseph Cummins, developed with Professor Hydén a method for measuring in single nerve cell bodies[4] the activity of the enzyme ATPase which breaks down ATP; these cell bodies had diameters of a fifteenth to a thirtieth of a millimeter; this was a very significant achievement, and we were able to modify the technique and find much more enzyme activity.[5] The experiment involved measuring very small concentrations of ATP, so I wondered if one could measure the change of this 'high-energy' compound in the retina (part of the eye), composed of thousands of cells, when light was shone on to it. I found a considerable change. Since the retina is an outgrowth of the brain, I sought the same effect in slices of brain, spinal cord, and the sciatic nerve (which runs down the back of the leg), and found that all these tissues exhibited it. Then I asked myself, "Why should one have a light-sensitive enzyme in the nerve upon which one sits?" After three months of intensive experiments; I left out all the tissue. To my astonishment, the ATP itself was sensitive to light. This was unexpected and had not been reported before. I repeated the experiments with a much *less* sensitive method of measuring phosphate.[6] The results were the same.

I have always been of the opinion that when a humble journeyman of a research worker finds something exciting about such an important molecule, it is likely to be either a mistake or an artifact resulting from the procedure. All living tissues employ complicated and fragile biochemical mechanisms. Most experiments involve killing an animal or plant, arresting change within it (fixation or inhibition), spinning it, freezing it, or adding powerful reagents. Any of these stages of a procedure can and often do change the biochemistry of the tissue drastically, or relocate particular chemicals within it. Therefore, biochemists have to demonstrate unequivocally that any effects they find arise from the innate properties of the tissues, rather than from the procedures used to examine them. Experiments to test the effects of the procedures themselves are known as 'control' observations, and the fundamental validity of any experiment designed to find out what happens in the intact human being, animal, or plant, is largely determined by the care with which the controls have been carried out. Popper[7] told us that one should try to falsify one's own hypothesis with relevant control experiments.

So I embarked on a long series of control experiments, testing the effects of light on compounds allied to ATP, including ADP, AMP, and inorganic phosphate; I tried the effect at 22°C rather than 37°C on ATP; I washed the glassware with detergents not containing phosphate; I took out the oxygen. None of the other substances showed the light sensitivity of ATP, which required oxygen, and occurred at body temperature, but not room temperature.

I made extensive literature searches and could not find previous reports of this finding. One day, my technician, Miss Anita

Bäckman, was away, and I was making up the solution. I took the bottle of ATP out of the refrigerator, and noticed that it was labeled, 'keep cool, in the dark.' So I wrote to Dr. Berger, the chief chemist of the manufacturer, Sigma, to ask if he knew of any publication of this phenomenon. He replied that his company had found it by accident, because it had been despatching chromatographically pure ATP from the United States, but customers in Europe had complained that they were receiving mixtures of ATP and its breakdown products, ADP and AMP. When Sigma put the ATP in dark bottles, it did not break down spontaneously. The company had not published the finding. However, I felt reassured by the knowledge that someone else had previously and quite independently detected ATP's sensitivity to light.

ATP used to be regarded as a 'high-energy' phosphate,[8] until the concept was reexamined[9] and it was also shown to be completely wrong by the redoubtable Dr. Barbara Banks, who had great difficulty in publishing these views.[10]

One day, Miss Anita Bäckman, Miss Inger Augustsson, and I were sitting in the warm room at 37°C with the lights turned off doing an experiment. It was necessary to study light sensitivity in the dark. We were talking to relieve the boredom of waiting for twenty minutes. That evening, when I analyzed the results of the experiment, it became clear that some agent in addition to light was having an effect. Virtually the only explanation was that the talking altered the stability of the ATP. I sought a source of sound, rich in high frequencies, and used some recorded bagpipe music. We started doing experiments early in the morning in the warm room, close to the entrance of the Institute. By coincidence, members of staff coming to work passed the door and heard the bagpipes, which they thought I was playing to the two young ladies. The bagpipe music had a considerable effect on the stability of the ATP, so we tested pure notes at different intensities, and different frequencies at the same intensity; they all showed that at 37°C, ATP was sensitive to sound. We subsequently showed that it was also sensitive to spinning in a desk top centrifuge at 1000 rpm for five minutes, electric current induced from a loudspeaker coil, and different concentrations of sodium and potassium ions in the range of concentrations normally found in the body.

ATP provides the immediate energy for muscle contraction,[11] but energy is stored in muscle as creatine phosphate,[12] and the 'high-energy phosphate' of cold-blooded animals is arginine phosphate.[13] So we repeated all the experiments we had carried out on ATP on creatine phosphate at 37°C, and arginine phosphate at 23°C. They all showed the same effects. Between 1962 and 1964, we did each experiment at least six times, and the analyses were made at random, the technicians not knowing when they measured the samples whether they were control or experimental, or when they had been extracted. We did about seventy experiments *before* we were satisfied that the technique[14] was as sensitive as we could make it, and then carried out over 330 further experiments for publication.

RETURN TO BRITAIN:
THE MEDICAL RESEARCH COUNCIL UNIT
OF APPLIED NEUROBIOLOGY, LONDON, 1964–1965

Light, sound, centrifugation, electric current, and physiological concentrations of sodium or potassium ions, all affected the stability of the three phosphates. Each of the six different kinds of energy could be transduced or converted into chemical energy for metabolism, so in 1964 I wrote a paper entitled, 'The phosphate bond as a transducer.' It was a tactical error to mention a hypothesis in the title, and also to include the experiments on bagpipes, as this enabled referees to trivialize them. I submitted the paper for publication to the *Journal of Physiology*, whose referees said that my reagents ATP, creatine phosphate, and arginine phosphate were not of the highest purity obtainable; I answered that such naturally occurring substances were not pure in living animals. The journal *Nature* said it had no room. The *Journal of Molecular Biology* gave no reason for rejecting it. The *Biochemical Journal* wrote to me that the idea "that physical agents could have biochemical effects was revolutionary." I replied that, on the contrary, it had been concluded that physical agents could have chemical effects, when Count Rumford at the time of the French Revolution showed that boring canons generated a great deal of heat. It became fairly clear that the journals did not wish to publish my paper. The referees did not like the findings, perhaps because they felt threatened by them, but I have never found out why.

Professor Hans Krebs—the Nobel Laureate in biochemistry—wrote to me that he thought a journal had the right to refuse to publish a paper if the referees *thought* that there was something wrong with it, but could not identify the error; I respectfully disagreed. Professor A. V. Hill—the Nobel Laureate in physiology—agreed that the effects had probably been demonstrated, but he could not recommend a journal which would publish the manuscript. Sir Ernest

Chain, whose method[15] I had modified, agreed that the modification was reasonable, and that I had, in fact, demonstrated the effects we claimed. Dr. Isaac Berenblum also agreed that we had modified his technique[16] suitably, but he would not comment on the experiments, as his field of research had moved to cancer.

In 1964, I presented my findings at the International Union of Biochemistry in Washington, the (British) Biochemical Society and the Physiological Society, in order to hear any new criticism, and to create a better climate for publication of a full paper. At the two biochemical meetings, the audiences made humorous or sarcastic comments, but very strange events occurred at the Physiological Society Meeting in Mill Hill, London, in November 1964.

About a fortnight before the meeting, Professor Max Born, of the Department of Pharmacology of the Royal College of Surgeons, asked me to come over to his laboratory to set up the ATP experiments which I was going to report. I had just done the first two experiments to set up the procedure. These were not accurate enough to give reliable results. He then told me that he had wasted enough time and he wanted to stop doing them. I told him that I did not think this was fair, since I had done over seventy experiments in Sweden before I was satisfied that the reliability was great enough to start a substantive series.

I was then the second most senior worker at the Medical Research Council Unit of Applied Neurobiology at the Institute of Neurology in London. The Director was Dr. John Cavanagh. He heard about my proposed paper, and said that someone—whom he refused to name—had told him that my paper would meet much opposition at the Physiological Society, and I would be wise to withdraw it. He would not say on what grounds it was to be attacked. His concern seemed so strong that I offered three times to withdraw it if he were to say that my presentation would damage the reputation of the newly formed unit. No, he insisted, that I had the right to present it, but still he advised me strongly against doing so.

Dr. Olof Lippold, the Reader in Physiology at University College, had agreed to introduce my paper, since I was not then a member of the Society. He also told me that I was going to be attacked by Professor Born, who sent me a summary of what he was going to say. I told Dr. Lippold that I believed that I had precise answers to any questions, including Professor Born's, which were likely to be raised.

Before the meeting, Dr. William Feldberg, the chairman of my session, said that he had heard that my presentation would be strongly attacked, and probably refused publication. This could happen by a simple majority of those who chose to vote, and would damage my career seriously. He even offered to say that I was not present, which would defer my paper to the next meeting. Knowing that Professor Born was to lead the attack, I offered to defer giving my paper if those who did not like it were prepared to try to repeat my experiments. I received no such undertaking.

As soon as I had finished my ten-minute presentation, Professor Born rose and showed one of the two experiments I had done in his laboratory. He asserted that since *these* two experiments had not shown a significant effect, those from Sweden I reported were not significant either. I answered that I did not see how only two experiments with 300% error could be used to invalidate about 330 with only 0.3% error. Professor G. S. Brindley said that I had not shown the effect of shining light on inorganic phosphate, or *not* shining light on ATP solution. I replied that these had been my first two slides. I was asked if I had used spectroscopically pure reagents. No, I answered, but the 'high-energy' phosphates were chromatographically pure. Had I tried the effect on ADP and AMP? Yes, I said, and the effect was not there, as I had said in my presentation. I was satisfied that I had answered every question fully and without equivocation.

I counted about two hundred people in the audience, of whom some were visitors. About four voted in favor of publication, about fifteen against—the rest abstained. The abstract was not published. Professor Born came up to me to say that he was sorry that my paper had been rejected. I did not answer him. Professor John Butler of the Chester Beatty Institute said that I would probably not now be able to find a job in physiology in Britain. My director, Dr. Cavanagh, said that he had heard that I had "made a fool of myself," and that the Medical Research Council did not like my research. I replied that before taking up the job in his unit, I had listed the experiments I wanted to do, including those on 'high-energy phosphates.' I requested to discuss this opinion about my experiments with those members of the Medical Research Council who did not like them, but he would not tell me their names or arrange a meeting.

The Physiological Society Meeting taught me something. Eventually I published the effect of light on ATP,[17] but not on creatine phosphate or arginine phosphate, nor the effect of light, sound, centrifugation, electric current, or sodium and potassium ions in the natural concentrations found in living tissues on creatine phosphate or arginine phosphate. However, a short report appeared,[18] and the full text was circulated by the Information Exchange.[19]

Some years earlier, the American, Dr. B. Chance,[20] the Russian Dr. S. E. Shnoll and collaborators,[21] and a Dutchman, Dr. F. A. Hommes,[22] had shown similar effects, so I wrote to each of them privately to ask if they had ever observed the phenomena I had seen. None of them answered, so I asked them through the Information Exchange,[23] but none of them replied. Their findings supported mine, but I still failed to obtain full publication in a refereed journal. My experiments remained suspended in an agnostic limbo, and my career was at risk.

SUBCELLULAR FRACTIONATION

In 1964, in Göteborg, I started looking into the theory of the effects of light, sound, electricity, and centrifugation. The latter was of particular interest, because centrifugation was so widely used in subcellular fractionation. Usually, when biochemists, biophysicists, cytologists, pharmacologists, or oncologists tell one what happens in, say, the nucleus, the cytoplasm or the membrane of cells, they have used the procedure of subcellular fractionation. This is intended to separate a fraction believed to be particularly rich in that part of the cell, so that its unique biochemical properties may be examined separately. The steps of the procedure are listed in the next paragraph. All experimental procedures used in the sciences imply necessarily the *assumption* either that the procedure itself does not change the biochemistry of the tissue being studied to a greater extent than the changes claimed between the control and the experimental tissues, or that any changes produced are too small to affect the result of the experiments. Does the final material extracted reflect the properties of the living tissue from which it came?

I used the following approach. I made a list of the steps of the procedure: for example, the animal is killed; it cools down; a particular tissue, such as the liver, is excised; a strong reagent is added to facilitate homogenization (mashing up); the tissue is cooled; it is homogenized; it is cooled again; the homogenate is centrifuged (spun round rapidly); fractions each believed to consist largely of a particular cell constituent are separated and frequently washed; substrate mixtures are added; the product is colored, so that the intensity of the color read on a spectrophotometer tells one the rate of a reaction in a particular part of the cells. Of course, there are numerous variations of this procedure.

Having identified the steps, I sought in the literature findings indicating the extent to which each of the steps of the procedure could change the properties of the part of the cell, its distribution, or activity. I then listed the assumptions built into the procedure, which had to be true if the properties of the cells were to reflect those in life. Finally, I listed the minimum control experiments which might satisfy one that measurements at the end of the long procedures reflected the original properties of the living intact structures.

The assumptions inherent in a procedure are crucially important, since, like a chain, the validity of a whole experiment is dependant on the strength of its weakest link. When I first examined subcellular fractionation,[24] I identified fifteen assumptions, some of them contrary to the laws of physics and thermodynamics; the second time I looked, I found twenty-four assumptions.[25] Other biochemists might deny that some of these assumptions were inherent, or they might add others, but, with one exception, they have not done so (please see page 111).

In 1972, I first raised the question of control experiments to test the effects of procedures on the final results of experiments.[26] It seemed that no one had done these, and this meant that the experiments were incomplete, and that conclusions could not be drawn from them, nor could theories be derived from the conclusions. My uncertainty about control experiments led me to write to the (British) *Biochemistry Society Bulletin*,[27] asking whether any biochemists knew of any published references that control experiments on the effects of the procedures on the results of experiments had indeed been done, or they would say that these were not necessary. There was no answer to these questions.

OTHER PROCEDURES WIDELY USED IN CYTOLOGICAL RESEARCH

I was so disturbed by the thought that subcellular fractionation might be an unsatisfactory technique that I decided to take a completely different technique and subject it to a similar analysis. I took electron microscopy, asking the question, 'How much does a picture taken with this instrument tell one about the structure of the living cell?' Since the early 1950s, there has been a passion for relating 'structure' to 'function,' that is, the appearance by electron microscopy of a particular identifiable part of a cell with the biochemistry it exhibits.

The light microscope had been used to examine living cells, unfixed tissue, and stained sections for one hundred years until the 1940s. At that time, the electron microscope was introduced. It permits much higher resolution and magnification than the light microscope, but the tissue can not survive the low pressure, the bombardment of electrons, and x-radiation in the electron microscope, so it has to be coated with a deposit of salts of osmium, lead, or tungsten, which is not destroyed by these agents, and can therefore be examined. Cytologists were very anxious to use this more powerful instrument to look at the fine structure of cells.

Science is so complex nowadays that frequently research workers have to resort to evidence derived by other specialists using techniques those citing them do not understand. They assume that their colleagues perform careful and valid experiments, whose fundaments have been examined adequately. My experience is that this is not always the case. I believe that a proper philosophy for scientists is that they should understand *all* techniques whose results they use or quote. They should be prepared to examine criticisms of findings they use as evidence in case their invalidity would throw doubt on the conclusions derived from them, and their value as evidence in other fields. Since truth should be universal, all scientists have a duty to resolve all anomalies and inconsistencies not only in their own beliefs but also in those they quote, and between their findings and those of other workers using the same and different techniques.

Unfortunately, electron microscopy was even more questionable than subcellular fractionation. So was histochemistry (the study of tissue sections), which I chose to analyze because it was somewhere between subcellular fractionation and electron microscopy. I then took three techniques nearer the measurement end rather than the preparation end of the former techniques: these were chromatography, electrophoresis, and radioactive measurements. Each of these had their burden of assumptions, many of them evidently unwarranted.[28]

For example, most cytologists know, but readers of elementary textbooks do not, that when one looks at an illustration of an electron micrograph: an animal has been killed; it cools down; its tissue is excised; the tissue is fixed (killed); it is stained with a heavy metal salt; it is dehydrated with increasing concentrations of alcohol; it shrinks; the alcohol is extracted with a fat solvent, propylene oxide; the latter is replaced by an epoxy resin; it hardens in a few days; sections one-tenth of a micrometer thick, or less, are cut; they are placed in the electron microscope, nearly all the air of which is pumped out; a beam of electrons at 10,000 volts to 3,000,000 volts is directed at it; some electrons strike a phosphorescent screen; the electron microscopists select the field and the magnification which show the features they wish to demonstrate; the image may be enhanced; photographs are taken; some are selected as evidence. One can immediately see how far the tissue has traveled from life to an illustration in a book.

I sought permission to do some of the control experiments, and was told that they would be a waste of time, 'controversial' and I would not be able to get the results published. Neither the Science Research Council, the Medical Research Council, nor the University of Surrey would support such a project. So I wrote a book about the uncertainty of biochemical techniques.[29] Well known publishers turned down my manuscript. The University of Surrey Press published it as its first book and printed twenty-five hundred copies. It sold out in Britain and the United States, but the publishers would not allow me to write a second edition. In 1975, unknown to me, the Russians translated it and sold out 12,500 copies.

The book was reviewed by *Nature*, the *Times Higher Education Supplement*, *Science Progress* and *Acta Biologica Academica Scientia Hungarica*. In the latter, Dr. Sandor Kerpel-Fronius wrote, "I feel strongly that the uncertainty is far from being as absolute as Dr. Hillman postulates. It should be remembered, for example, that in situ experiments with radioactive tracers and those carried out in vitro on isolated organelles or enzymes are very often in essential harmony in spite of the unquestionable shortcomings of the methods used. Such corroborative results should assure us that the present understanding of at least some biochemical processes is close to reality."

At the time, I regarded this as the substantive answer to my reservations about biochemical techniques. While each technique only gave an approximation to the whole picture, the whole story put together produced a consistent picture.

The most hostile reviewer of my book was Professor J. Lucy of the Royal Free Medical School. He wrote, in *Biochemical Education*,[30] that I had overstated my case "in stating that the validity of a localisation of an enzyme activity is dependant upon all fifteen assumptions listed being warranted," then he added "(sic)." This implied clearly that he himself did not believe that the conclusion of an experiment *must* depend upon all its assumptions being warranted—he was saying this to lecturers, teachers, and students. He pointed out a real mistake I had made about liquid scintillation although it did not affect my argument.

LIGHT MICROSCOPY AND THE BRAIN,
UNIVERSITY OF SURREY, 1965 TO DATE

Mr. Peter Sartory, although an amateur, was in my judgement one of the most expert British light microscopists. He was a distinguished natural historian, a former microscope manufacturer, an amateur astronomer, a former Committee Member of the Royal Microscopical Society, and former President of the Quekett Microscopical Club, founded in 1865. By the time I met him in 1967, he was chronically ill with lung disease, having smoked heavily all his life.

I had recently returned from Sweden, and asked Mr. Sartory if he was interested in looking at single fresh mammalian nerve cells dissected out from the brain by the technique which Hydén had used to examine the properties of cells.[31] Hydén himself had agreed that it would be useful to look at these cells by a variety of light microscopical techniques in the unfixed state, which is the nearest to the living state in which cells can be examined.

Hydén always took the cells out in a sugar solution,[32] but we tried taking them out in saline, which was more natural. Immediately, we saw a membrane around the nucleolus,[33] which had not been seen before. We tried to publish this finding, even resorting to the very ancient practice of sending the editor of various journals not only the photographs, but slides of cells showing our membranes. We demonstrated it to the distinguished microscopist, Dr. John Baker, who agreed that he could see it. The journals turned it down, firstly, because it had not been shown by electron microscopy, which distorted it;[34] secondly, it had not been seen before; thirdly, we had not shown that our membrane consisted of lipids and proteins, as the Davson-Danielli and the Singer-Nicolson models assumed. We pointed out that any lipids in the original membrane would not have survived the extraction by alcohol and propylene oxide during preparation for electron microscopy—despite all textbooks of life sciences showing cell membranes they believe to be of this composition. So far, there has been a Trappist response to this rather awkward point by the many electron microscopists with whom we have tried to discuss it.

I have never understood the reasons for resistance to the belief in the nucleolar membrane, even although we have published many photographs of it[35] and offered to send microscope slides to anyone in the world who wanted to see it.

ELECTRON MICROSCOPY

Although at that time we did entertain doubts about the value of electron microscopy in biology of tissues containing much water, both of us still felt that the best source of information about the fine structure of cells was probably electron micrographs. We were comparing our high contrast light micrographs of unstained nerve cell bodies with the latest electron micrographs. We suddenly noticed that the endoplasmic reticulum, which is a network believed to be a structure in the cytoplasm (the cell sap), appeared to be cut perpendicular to the section far too often than solid geometry would permit. It was as if one threw into the air a large number of coins, and when one photographed them, the vast majority were to appear edge on—instead of in all possible orientations. Unfortunately, this was also true of all the apparent membranes in the cell, the Golgi apparatus, the mitochondrial membranes, the cell membrane, and the nuclear membrane.

Whereas we did not doubt the existence of the cell, nuclear and mitochondrial membranes, their sandwich ('trilaminar') appearance was simply impossible in solid geometry (fig. 5.1). Obviously, if they were random, they should be seen as flat sheets as often as they are seen in almost perfect transverse section. After we had begun to doubt the existence of the cytoplasmic network (the endoplasmic reticulum) and the Golgi body on *geometrical* grounds, we suddenly realized that if they existed, they would not permit the intracellular movements which are generally regarded as evidence of the life of the cell. These movements can be seen by low power light microscopy, while there is supposed to be a fine network throughout the cytoplasm requiring very much higher magnification to see. Furthermore, iron filings, carbon particles, or pollen injected into the cytoplasm spread quite freely, quite unimpeded by any fine network which would shackle them.

Thus we had two quite different lines of evidence—each of them powerful enough to question the existence of all the new structures in the cytoplasm seen with the electron microscope, plus the Golgi apparatus. When we were satisfied that we had overwhelming evidence, we submitted a paper for publication. The editor of *Nature* rejected it on the grounds that the referee believed that no one at that time (1975) still believed in the 'unit membrane' or that the reticulum was attached to the cell membrane or nuclear membrane, but added that he agreed with us. *Science* gave no reason for rejection.

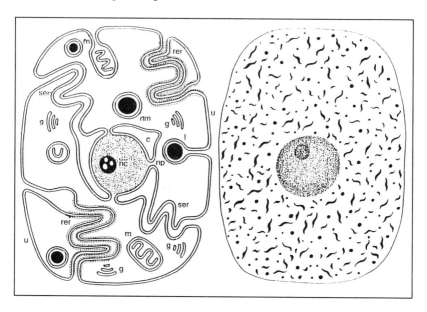

FIGURE 5.1. The structure of the cell, at the left as agreed by most modern cytologists and at the right as believed by me. In the structure on the left, *u* is adjacent to the double cell membrane, *g* to the Golgi apparatus, and *ser* and *rer* to the endoplasmic reticulum, a network in the cytoplasm; *m* is a mitochondrion containing the 'shelves' of cristae; *np* represents holes in the nucleus, the 'nuclear pores.' In my publications, I have shown that the double cell membrane should not always appear to be cut at right angles, and the reticulum or network would prevent intracellular movements which are characteristic of living cells. In the structure on the right, the mitochondria appear in the cytoplasm smaller and are oriented randomly. Further details are given in references.[70]

Scientific American would not consider it as it had not been previously published; after it had, they still would not accept a small piece about our views. Nor would *La Recherche* or *New Scientist*. The latter wrote to me that it did not accept controversial articles, in a letter I received the same week as it featured Dr. Rupert Sheldrake on the exceedingly controversial concept of 'morphic resonance.'

As for the assertion that no one then believed in the 'unit membrane' or that it was attached to the cell or nuclear membrane, I listed all the latest books on life sciences in the reference collection of the University of Surrey; every single one of them indicated that they believed the former points. At meetings of learned societies, whenever anyone alleged that these beliefs were no longer agreed, I produced photocopies of these lists, and challenged those who had denied my assertion to name one textbook or paper (other than our own) which said that it did *not* believe in the 'unit membrane' or the attachment of the reticulum to the cell and nuclear membrane. Subsequently, the cytologists replied, "You don't want to believe what you read in textbooks." We were so horrified by this latter sentiment that we wrote a letter to *Nature* asking anyone at the Physiological Society, the Anatomical Society, or the Royal Microscopical Society to justify this view in writing.[36] No one replied. So we put this list in a book we subsequently wrote on cell structure.[37]

Let us be clear what was happening. Firstly, senior research workers recommended only textbooks containing description of cell structures with which they disagreed; then they denied that these structures were so described; then they could not name any books or papers describing what they taught as the correct cell structure; then they alleged that one should not believe what one read in textbooks; then they were not prepared to justify such cynicism in print. The situation has not changed.

Most cell biologists today believe that intracytoplasmic movements of subcellular organelles occur, but also that there is a fine dense cytoskeleton in the cytoplasm; they believe that the image of the cell seen by electron microscopy is three-dimensional, but have to tilt the stage to show some of the orientations. Thus, the views they hold about cell structure are inconsistent. This is extremely worrying, because the use of so many approximate techniques in biochemistry is usually justified by the assertion that together they make a fully consistent story (please see page 111). If they do not, the justification for using such popular techniques becomes even weaker, and the urgency to examine the validity of the procedures even greater.

We were quite unable to obtain publication of our paper showing that all the structures in the cell first shown by electron microscopy *plus* the Golgi body were artifacts. Among the criticism we faced was that we were not electron microscopists, although I had been using the instrument for seventeen years when this was first said. Even if it were true, we believe that we do have a right to use the currency which authoritative electron microscopists have put into circulation.

Another tactic used against us was to label our ideas as 'old hat.' Critics said that ideas had been considered in the 1940s and 1950s, when the electron microscope was first used for biological tissues, and refuted then. Unfortunately, they could produce no references to support this assertion. Another trick was to say that all biochemists and cytologists studied artifacts. This was a smart way of avoiding discussions of which artifacts gave useful information about cells and which did not. A social trick was to exaggerate or joke about our views. Said one chairman, "As you know, Dr. Hillman does not believe in membranes—ha-ha!." At a coffee queue at a Physiological Society Meeting at University College, London, I heard two students agreeing that "Hillman's views are rubbish." "Have you ever read any of his papers?" I asked. "Of course not, I would not waste my time."

Among the more difficult questions raised by our critics were 'What do you mean by truth?,' 'What is an artifact?,' 'When is it useful?,' 'What do you see when you look down a light microscope?,' 'What is the nature of the image seen by the electron microscope?' It is only fair to say that these questions are not usually raised when one submits a paper on the structure of cells for publication. Various well-meaning, but perhaps naive, friends suggested that if we expanded our manuscript into a book, also dealing with these fundamental questions, we might have a greater chance of publishing it.

With the Audio Visual Aids Unit of the University of Surrey, we made a thirty-five minute film, and I showed it at the International Physiological Society Meeting in Paris, the Biochemical Society in Cambridge, the Society of Experimental Biology at Brighton, the Quekett Microscopical Club in London, the Physiological Society at University College, London, and other places.

At University College, Dr. A. Lieberman, the well-known electron microscopist, said after the film had been shown that he had many pictures of the endoplasmic reticulum and the 'unit membranes' in all orientations, which we had denied. Since his laboratory was close at hand, I suggested that he go to get them immediately to show the audience. His laboratory was untidy, he said, so I asked him if he would show me these micrographs if I called on him. I telephoned him five times altogether, but he would not send me any micrographs or references showing the images I requested. He did, however, send me an interesting reference on interpreting electron micrographs, which I did not feel was a relevant response. I offered to announce it in public, if I were to receive a micrograph showing a full range in the expected incidence of any of the structures whose existence we had doubted, but the full range would have to be in the *same picture*. That offer still remains open.

In 1977, I was invited to give a forty minute paper at the 1979 Leopoldina Symposium on 'Cell Structure' in Thuringia, Germany, but when I arrived, I found that I was not on the program. The Secretariat would not accept my paper for publication on the grounds that it had not been received beforehand, although I had sent in the manuscript over six months before. I offered the Secretary another copy, but was told that it was too late. As I was waiting, I heard the Secretary telling a young research worker, who apologized for bringing her manuscript with her, that there was "plenty of time." When I pointed this out, the Secretary became confused, and referred me to the organizers. Nevertheless, by dint of diplomacy, I showed my thirty-five minute film, and it was followed by a fifty minute discussion. My paper was not included in the proceedings.

We had written a paper which we could not get published, a book that solicitous referees were sure would be published by another publisher, and a film which was irritating audiences. One day, I was invited to show our film by the Bristol Fine Structures Group, which was composed mainly of electron microscopists. Professor Richard Gregory, the psychologist, was in the audience, and asked me if we had ever tried to get our views published. Yes, I said, but without success. Had we thought of *Perception*, the journal of which he was editor? I said that our geometrical points were suitable for his journal, but not the biological ones. He asked me if I would be prepared to rewrite the paper to emphasize the geometrical arguments. Mr. Sartory and I were rather reluctant to do this, and then have the paper rejected for publication. Eventually we agreed to rewrite it and submit it, on condition that we did not have to leave out any points of substance. It was eventually published,[38] and the editor told me that "I have had a lot of stick from the electron microscopists for doing so."

About July 1977, Mr. Sartory suggested that we had to open a 'second front'—that was, we had to tell the *public* about this situation. I was reluctant, and it took him three months to persuade me that we had properly explored all the usual scientific channels. He telephoned the London *Observer*, and I was interviewed by its then science correspondent, Mr. Nigel Hawkes. He kept trying to discuss the economic consequences of the enormous cost of electron microscopy in medical research, and I kept trying to talk about the scientific points. About three months later, a short article appeared on the front page.[39] Exactly a week later, the *Observer* published a letter from the senior elected officers of the Royal Microscopical Society.[40] They said that our views had been generally rejected "in the light of overwhelming evidence to the contrary." They said that biologists had not seen our views in print. (We thought that this was somewhat cynical, since some of the signatories of the letter had resisted the publication of our manuscript).

Two weeks later, we replied.[41] "We know of no circumstances in which our views have been generally rejected. . . . Nor do we believe that in scientific discussions, the correctness of an idea is measured by the number of its supporters." We ended, "The protagonists of the current view have so far been remarkably reticent in discussing with us these important questions. May we, Sir, through the hospitality of your columns invite the distinguished signatories of the letter, or anyone else who agrees with them, to debate these questions in front of a scientific audience at any place or at any time."

Two years later, although I had lectured widely, no one had responded to our invitation, so we repeated it in the *Observer*.[42] One debate eventually occurred, Dr. John Douglas of Brunel University arranged it with Dr. A. Robards of York University and the Royal Microscopical Society and Dr. K. Roberts of the John Innes Institute on the one hand, and Mr. Peter Sartory and myself on the other. By that time, Mr. Sartory was so ill that he only spoke for a few minutes. Dr. Robards started by saying that he was speaking in a private capacity, not representing any organization. The new point he made in reply to the question about why most of the membranes appeared edge on in electron micrographs was that, like a barn door, one would not necessarily see it if it were open. I replied by asking why one did not see a *space* corresponding to the size of the door when it was closed. Mr. Anthony Tucker, the science correspondent of the *Guardian*, gave a short summary of the meeting,[43] and Dr. P. Evennett, a signatory of the letter attacking us,[44] wrote an account in the *Proceedings of the Royal Microscopical Society*.[45]

We wrote to Sir Peter Medawar, the Nobel Prize Laureate who has published several books of advice on 'good' science, telling him that we would like to see him, and enclosing a four page summary of our views; we pointed out that these would be understandable to a student aged fifteen. He replied that he could not comment as he was not an electron microscopist, but he referred our letter to 'his' electron microscopist. Five polite letters and ten years later, we have not heard from 'his' man. A recently knighted Oxford professor invited me to discuss the matters; I traveled all the way from Guildford (taking about four hours each way), but he only had twenty minutes to see me, not enough time, I suppose, to comment on what I said or offer me a cup of coffee. He passed me to his electron microscopist, who, after fifteen minutes, said, "I have to go to lunch."

A famous textbook writer agreed that the Robertson model of the 'unit membrane'[46] was impossible, but said that he could not take it out of the next edition. When I asked him why not, he smiled benignly. At about that time, the publisher Mr. Michael Packard agreed to publish our book,[47] in the belief, which we shared, that biologists in general would be very interested in a new look at the structure of the living cell.

As I traveled round, there was hardly a single place at which I lectured where several people did not come up to me when I had finished and say that they agreed with us. I always asked them their names. Would they be prepared to say in public that, for example, they did not think the cell membrane was trilaminar (fig. 5.1) or that there was a reticulum in the cytoplasm? (Mentioning our names was not necessary.) One lecturer in Edinburgh, whose name I did not note, said he would. All the others had reasons why they could not: they were writing theses, seeking lectureships, applying for grants, or being considered for chairs. Did I blame them? Do you? What would happen to their careers if they embraced controversial views?

In 1981, not long afterwards, BBC Television made a 'Horizon' program, 'No one will listen to me.' It took the cases of Professor John Laithwaite, Dr. John Hastead, and ours. A fair attempt was made to summarize our views, but these were 'answered' anonymously by the program—those disagreeing with us not appearing. Of course, this gave the impression that our views were so ridiculous that the 'Establishment' did not itself want to counter them. Needless to say, we had answered the particular points several times. Although the producers of the program did not intend this, the effect was most unfair to us. Nevertheless, the program no doubt gave our views more visibility than desired by our critics.

Soon after the broadcast, I was taken off all undergraduate teaching in the University of Surrey, without any reason being given. I was an elected member of the Senate. So I asked at a meeting attended by the Dean and the Heads of the Biological Departments whether I had been removed from the teaching because of my views on cytology. I pointed out that I was the senior physiologist, that I taught relatively little cytology, and that I told my students the accepted wisdom because I wanted them to pass their examinations. There was no answer. The Department of Human Biology was closed down a couple of years later. Although I was the second senior person in the department and had London University degrees in medicine, physiology, and biochemistry, I was the only member of that department not placed elsewhere in the University.

Our policy in the 1970s had been to refuse to speak to school or undergraduate students, because doubts at that stage might discourage them from learning biology at all. However, we changed our minds, firstly because we realized that ten to sixteen year olds were being taught about the reality of structures we said were artifacts at a time when they believed that everything they learnt was gospel truth. They were so conditioned. Secondly, we were persuaded that motivated young people often learned better when they were presented with opposing views. Thirdly, even the briefest acquaintance with the history of science teaches us that advances have nearly always been made when established views were reexamined.

Mr. A. Bishop, the Editor of *School Science Review*, which was widely read by school science teachers, invited us to submit a manuscript. We wrote an eleven page paper.[48] Two groups of electron microscopists, Dr. R. W. Horne and Dr. J. R. Harris,[49] and then Dr. R. H. Michell, Dr. J. B. Finean and Dr. A. Coleman[50] took issue with us in writing; they acknowledged the help of Professor W. E. Coslett, Dr. A. W. Robards, Dr. J. Burgess, Dr. S. Hunt, and Professor H. W. Woolhouse. They attacked us, inter alia, for not having dealt with the considerable volume of data from biochemistry which they believed supported their view. Although we had cited it several times in our paper, they had failed to notice that our views had originated from a book I had written about just this subject.[51] We wanted to reply to the points made by the electron microscopists. Our original paper had been twelve pages long, and theirs together added up to twenty-five pages, but we were only accorded a letter of five pages to reply. We therefore decided to list the eleven questions which we had raised[52] which had not been answered. We wrote to all the authors of the two papers and those whom they had acknowledged, inviting them to an informal discussion of the differences between us. Only Dr. Michell replied, and he was not willing to discuss these matters with us. He was subsequently elected a Fellow of the Royal Society.

This brings up some fundamental questions about the behavior of scientists. Do they have a duty to engage in serious dialogue about their published work? Is it satisfactory that they should not answer letters? What should students think about this? I subsequently published a full-length paper addressing all the points our electron microscopic colleagues had ever raised in discussion with us or in publication.[53]

STRUCTURE OF THE BRAIN

Every day pathologists examine beautifully stained thin sections of brain. They see the nerve cells and occasional nuclei clearly but most of the section does not stain at all, when viewed by light microscopy. In 1846, the great German histologist Virchow gave the name 'neuroglia' or 'nerve glue' to this unstained material.[54] The general consensus among neurobiologists today is that there are four kinds of cells in the brain and spinal cord, besides the blood vessels. The nerve cells are the excitable cells, which show up; the nuclei belong to the other cells, the neuroglial cells, classified into astrocytes, oligodendrocytes, and microglia. The neuroglial material is believed to consist of the three latter types of cells, with very little space in between them.

After I had been taking out nerve cells for about seventeen years, by Hydén's technique,[55] I conceived of the idea that Virchow was right—the unstained material in the brain was *not* composed of neuroglial *cells*.[56] I did a series of experiments lasting another five years, and they all supported the following conclusions. There are relatively few, widely spaced nerve cells in the brain and spinal cord. Any cell with processes (like wires) is a nerve cell. The greatest proportion of the central nervous system is a ground substance consisting of a fine granular material with 'naked nuclei.' I published this conclusion in a monograph[57] containing much evidence from the literature as well as my own experiments. It contained blurred micrographs, was camera-ready, and was expensive. Dr. J. R. Parker reviewed it in a neutral fashion in the *Lancet* and Professor Brian Leonard was laudatory in *Neurochemistry International*, but it sold badly.

I was also unwise enough to find that the evidence for the existence of the synapses—by which nerve cells are believed to communicate—contains so many inconsistencies, that they are likely to be staining artifacts.[58] I analyzed transmission, whereby signals are believed to pass from one part of the nervous system to another, and concluded that the view that transmission was chemical, formulated in detail by Professor Sir Bernard Katz,[59] contained too many unproved and unprovable assumptions for the theory as a whole to be acceptable.[60] I have also spoken about this at many meetings, but so far no one has addressed my objections to the theory. However, I felt a duty to propose an alternative hypothesis.[61] Of course, there is not enough room here to give evidence for these conclusions, but they are given in detail in the *published* references cited.

CLOSURE OF THE UNITY LABORATORY

In 1988, the Vice-Chancellor of the University of Surrey forced me to take 'voluntary' early retirement, on the following grounds.

- The University was short of money. (It has since acquired enough to set up five research professorships, and will receive £30,000,000 from its Research Park by 2000.)
- The University had selected those areas which it wanted to support, but mine was not among them. (I have never been able to persuade the University to tell me [1] what committee met, [2] whom else it considered to select my work for not supporting, [3] why I was not asked to submit my publications or an annual report of my laboratory's work, or [4] why my laboratory's work was *not* submitted to the University Grants Committee for evaluation.)
- My work was of poor quality. (I had published at least eighty full-length papers, mostly in refereed journals, and I had written three books by 1988.)
- I had not obtained outside funds. (Nor had about 70 percent of the academic staff.)

The Senate, on 30 September 1987, approved a 'Revised Academic Plan' for 1987–1990,[62] in which all departments were cut by 5 percent, but my laboratory was to be cut by 100 percent, that is, closed. This was approved by the Council of the University on 18 December 1987.

I took drastic action. On 5 November, I presented a paper to the Finance and General Purposes Committee showing that my laboratory was the cheapest in the faculty, that most academic staff members of the University of Surrey did not have outside funds, and that I had an above average research output. At least twelve of my senior friends, mostly from abroad, wrote to the Vice-Chancellor supporting me, although he did not report this to Senate or Council. Articles about the proposed closure appeared in the *Times*, the *Guardian* and the *Times Higher Education Supplement*. Two resolutions opposing the closure of the Unity Laboratory were passed unanimously at the Annual Council of the Association of University Teachers in 1988. A question was asked in Parliament.

The Handicapped Children's Aid Committee of London, which founded the Unity Laboratory and financed it from 1968–1981, rallied around and promised me support for one year. Dr. David Horrobin, Managing Director of Scotia Pharmaceuticals, who had himself suffered for his scientific views in Canada, also came to my aid. With this outside funding, the University agreed to allow my laboratory to remain open for a further three years, but without any support from University funds for my research.

'VOLUNTARY' RETIREMENT

Not long afterwards, I was asked to take 'voluntary' early retirement. The University offered to buy in seven years of my pension, to give me a lump sum, and to reengage me for 40 percent part time. The total financial settlement would leave me with almost as much income as if I were still a full-time assistant professor, but I would lose my tenure. I was given three two-day ultimata delivered by the hand of an Assistant Secretary of the University.

The Association of University Teachers took legal advice in my support. Our University had one of the strongest tenures in the country. I had thought that I was fully protected. However, the Association advised me that if the University dismissed me illegally, I would have to take legal action against it. If I won, damages to me would only relate to my loss of income, *not* my senior position, research facilities, prestige, and so forth. Such a case had not been heard before. The certainty of my winning was not by any means absolute. The consequences of failure of my plea would be financially disastrous to me, and I had a wife and four young children to support. The Association advised me to take the offer.

Reluctantly, I agreed to surrender my tenure under a number of conditions, not all of which were met. I believe that I am the only tenured academic in Britain who has lost his tenure because of his or her scientific views. Strangely enough, a few months before, I had ended an article on academic freedom in the *Times Higher Education Supplement* with the sentence, "Would you not be thankful that you had tenure, and lived in a democratic country?"

I have continued my full time research work with my colleague Mr. David Jarman in the Unity Laboratory. We have produced an atlas of the human nervous system,[63] and I have written a book, originally entitled *Letter to Students of Biology of the Twenty First Century*, now with a new name.[64] I have also listed the mechanisms whereby the dissemination of unpopular views is prevented in liberal societies.[65]

In recent years, without any reason being given, I have been prevented from presenting my views at a joint meeting in Würzburg of the German and British Physiological Societies (they told me that mine was the only paper they would not allow to be presented). The European Society of Neurochemistry would not allow me to speak at Leipzig. The British Society of Neuropathology prevented me showing a film because the Society said that it had seen it before—remarkable, because it had never been shown before. The joint meeting of the Norwegian and British Biochemical Societies at Eidsvoll invited me to send in an abstract and then would not publish it; they said that it was only because the film could not be understood unless one saw it, *but* they would not tell me how many others had been refused publication. At a meeting in August 1992 of the European Society of Neurochemistry in Dublin, although I was a founder member, I was speaking on a subject relevant to most other papers, and had requested an oral presentation, I was given the last slot at 5:15 p.m., after 171 papers at the end of a five day meeting. The chairperson did not turn up, the room was changed, several speakers did not arrive, and several others who wanted to hear my talk missed it. I received an apology, but no redress.

PRESENT SITUATION

I have shown, to my own satisfaction that (1) at least some popular important biochemical research techniques have never been controlled, (2) most of the new structures in cells apparent by electron microscopy are artifacts, (3) there are only nerve cells and naked nuclei in a ground substance in the brain and spinal cord, (4) there are no synapses, (5) the transmitter hypothesis is doubtful. I have published all the evidence for these statements, although this has not always been easy.

The stakes are high. If I am right a very large proportion of experiments in basic research in life sciences will have to be completed, and this may result in quite different conclusions. If I am wrong, only my reputation is destroyed. It would be natural for a lay person to think that it would be very unlikely that any single individual was right and nearly all other life scientists wrong. Even if my conclusions were totally correct, it is very unlikely that they would be acted upon, because there are so many academics, doctors, teachers, and publishers who have a vested interest in current views. History tells us that this does not happen quickly.

Every day that goes by more people have carried out more experiments apparently compatible with the current consensus, therefore more people have a career interest in it being correct. At the same time, in Britain at least—where academic tenure has been virtually abolished—it is unlikely that anyone who raised the fundamental questions or came to the same conclusions publicly as Mr. Sartory and I have, would ever be appointed to a lectureship, be awarded a large grant for research, or enjoy a successful career in science.

There is a widespread belief that medical and biological research is very successful[66] and therefore, more resources should be put into it. I have differentiated between two aspects of medical research. Since the 1940s, many new drugs have been discovered and developed empirically, intensive care units for dying patients have been set up in most large towns, new antibiotics have been found empirically and modified, transplantation of skin, kidneys, and other organs has become routine, cardiac surgery has become a major speciality, and steroids have been used for skin diseases. All these have been highly successful applications of simple *technologies*. However, we must ask what has been discovered about the *genesis* of cancer, multiple sclerosis, Alzheimer's disease, or schizophrenia. The answer is remarkably little which has helped us to understand the *mechanism* of the diseases, so that we can design rational treatments for them. The same may be said about the understanding of the molecular mechanisms by which drugs act; a large amount is known about *what* they do, but remarkably little about how they act in the living person or animal.

If we leave aside my hypothesis that basic medical, biological, and pharmacological research has not been successful because it has not addressed the fundamental problems and assumptions inherent in most of the techniques, the current situation is dangerous because it suppresses free thought, without which the advance of knowledge can only be slow.

MESSAGE FOR THE FUTURE

Irrespective of the truth or otherwise of my views in biology, I believe that it would be generally agreed that there is an international tendency to increases in: size of research units; complexity of research; cost of carrying it out; competition for academic positions; power of those who decide on the allocation of research funds; influence of those who control prestigious research journals; and censorship by the establishments of access to the popular media. It would also be agreed that knowledge can only advance when the current consensus is challenged. This is usually a consequence of thought by one or a few individuals, who by definition constitute a minority. Thus it is reasonable to be concerned that current trends will increase conformity and decrease individual or minority challenges, which will slow down the advance of knowledge.[67]

In addition, the large number of mechanisms discouraging the dissemination of challenging and new ideas will discourage intellectual honesty,[68] which is the overwhelming force which advances knowledge. Thus, the present situation will discourage academics from free thought. I would like to give a historical warning to all biologists that, unless they address some of the fundamental questions which I have raised[69] they are in danger of spending the whole of their research careers, using thermodynamically illegal procedures, studying artifacts, repeating uncontrolled experiments,[70] indulging in intellectual casuistry, or becoming cynical—none of which is good for science.

NOTES

1. H. Hydén, "Quantitative assay of compounds in isolated fresh nerve cells and glial cells from control and stimulated animals," *Nature,* vol. 184 (1959): 433–435.

2. H. Hillman, "Hydén's technique of isolating mammalian cerebral neurons by hand dissection," *Microscopy,* vol. 35 (1986): 382–389.

3. H. Hillman, "Fraud versus carelessness" (letter), *Science,* vol. 326 (1987): 736.

4. J. T. Cummins and H. Hydén, "Adenosine triphosphatase in neurons, glia and neuronal membranes of the vestibular nucleus," *Biochimica et Biophysica Acta,* vol. 60 (1962): 271–283.

5. H. Hillman and H. Hydén, "Characteristics of the ATPase activity of isolated neurons of rabbit," *Histochemie,* vol. 4 (1965): 446–450.

6. I. Berenblum and E. Chain, "Studies on the colorimetric determination of phosphates," *Biochemical Journal,* vol. 32 (1938): 286–298.

7. K. Popper, *Objective Knowledge,* Oxford: Clarendon Press (1972): 14–17.

8. F. Lipmann, "Metabolic generation and utilization of phosphate bond energy," *Advances in Enzymology,* vol. 1 (1941): 99–162.

9. P. George and R. J. Rutman, "The 'high-energy phosphate' bond concept," *Progress in Biophysics and Biophysical Chemistry,* vol. 10 (1960): 1–53.

10. B. Banks, "A misapplication of chemistry in biology," *School Science Review,* vol. 179 (1970): 286–297.

11. E. W. Taylor, "Chemistry of muscle contraction," *Annual Review of Biochemistry,* vol. 41 (1972): 577–616.

12. K. Lohmann, "Uber der enzymatische Aufspaltung der Kreatinphosphorsaure; zugleich ein Beitrag zum Chemismus der Muskelkontraktion," *Biochemische Zeitschrift,* vol. 271 (1934): 264–277.

13. J. F. Morrison and A. H. Ennor, "N-phosphorylated guanidines," in P. D. Boyer, H. Lardy, and K. Myrbäck (eds.), *The Enzymes,* vol. 2, New York: Academic Press (1960): 89–109.

14. Hillman and Hydén, op. cit.

15. Berenblum and Chain, op. cit.

16. Ibid.

17. H. Hillman, "The effect of visible light on ATP in solution," *Life Sciences,* vol. 5 (1966): 589–605.

18. H. Hillman, "The terminal phosphate bond as a transducer" (abstract), *Sixth International Congress of Biochemistry,* New York City (26 July–1 August 1964).

19. H. Hillman, "The phosphate bond as a transducer," *Information Exchange No 1,* vol. 190 (3 July 1964): 1–33.

20. B. Chance, "Spectra and reaction kinetics of respiratory pigments of homogenised and intact cells," *Nature,* vol. 169 (1952): 215–221.

21. S. E. Shnoll, M. N. Kondrashova, and K. F. Sholts, "Multi-phase alterations of the ATPase activity of actomyosin preparations and the effect of different features," *Voprosy Meditsinskoi Khimii,* vol. 3 (1957): 54–63.

22. F. A. Hommes, "Oscillation times of the oscillatory reduction of pyridine nucleotides during aerobic glycolysis in brewer's yeast," *Archives of Biochemistry and Biophysics,* vol. 108 (1964): 500–504.

23. H. Hillman, "Do oscillations in biological system originate in their substrates and nucleotides?" *Information Exchange No. 1*, vol. 298 (14 January 1965): 1–2.

24. H. Hillman, *Certainty and Uncertainty in Biochemical Techniques*, Henley on Thames: Surrey University Press (1972): 34–35.

25. H. Hillman, "Some fundamental theoretical and practical problems associated with neurochemical techniques in mammalian studies," *Neurochemistry International*, vol. 5 (1983): 1–13.

26. Hillman (1972), op. cit.

27. H. Hillman, "Control experiments for biochemical techniques," *Biochemical Society Bulletin*, vol. 1, part 3 (1979): 11.

28. Hillman (1972), op. cit.

29. Ibid.

30. J. A. Lucy, "Review of *Certainty and Uncertainty in Biochemical Techniques*," *Biochemical Education*, vol. 1 (1973): 32.

31. Hydén, op. cit.; H. Hydén, "The neuron," in J. Brachet and A. E. Mirsky, eds., *The Cell*, vol. 4, New York: Academic Press (1961): 215–324.

32. Hydén (1959), op. cit.

33. T. S. Hussain, H. Hillman, and P. Sartory, "A nucleolar membrane in neurons," *Microscopy*, vol. 32 (1974): 348–353.

34. I. Chughtai, H. Hillman, and D. Jarman, "The effect of haematoxylin and eosin, Palmgren's and osmic acid procedures on the dimensions and appearance of isolated rabbit medullary neurons," *Microscopy*, vol. 35 (1987): 652–659.

35. Hussain et al., op. cit.; H. Hillman and D. Jarman, *Atlas of the Cellular Structure of the Human Nervous System*, London: Academic Press (1991).

36. H. Hillman and P. Sartory, "Value of textbooks" (letter), *Nature*, vol. 267 (1977): 102.

37. H. Hillman and P. Sartory, *The Living Cell*, Chichester: Packard (1980): 102–105.

38. H. Hillman and P. Sartory, "The 'unit' membrane, the endoplasmic reticulum, and the nuclear pores are artifacts," *Perception*, vol. 6 (1977): 667–673.

39. N. Hawkes, "Mirages from microscopes," *Observer* (11 December 1977).

40. A. W. Robards et al., "Whose illusions" (letter), *Observer* (18 December 1977).

41. H. Hillman and P. Sartory, "Whose illusions" (letter), *Observer* (1 January 1978).

42. H. Hillman and P. Sartory, "A challenge repeated" (letter), *Observer* (16 December 1979).

43. A. Tucker, "A sight too powerful," *Guardian* (22 May 1980), and "Seeing is believing," *Guardian* (5 June 1980).

44. Robards et al., op. cit.

45. P. J. Evennett, "Fact or artifact. Debate on the reality of the unit membrane," *Proceedings of the Royal Microscopical Society*, vol. 15 (1980): 334–335.

46. J. D. Robertson, "The ultrastructure of the cell membranes and their derivatives," *Biochemical Society Symposia*, vol. 18 (1959): 3–43.

47. Hillman and Sartory (1980), op. cit.

48. H. Hillman and P. Sartory, "A reexamination of the structure of the living cell and its implications for biological education," *School Science Review*, vol. 61 (1980): 241–252.

49. R. W. Horne and J. R. Harris, "The electron microscope in biology," *School Science Review*, vol. 63 (1981): 53–69.

50. R. H. Michell, J. B. Finean, and A. Coleman, "Widely accepted modern views of cell structure are fundamentally correct," *School Science Review*, vol. 63 (1982): 434–441.

51. Hillman (1972), op. cit.

52. H. Hillman, "Some questions not answered by the electron microscopists" (letter), *School Science Review*, vol. 224 (1982): 173–177.

53. H. Hillman, "Some microscopic considerations about cell structure. Light versus electron microscopy," *Microscopy*, vol. 36 (1991): 557–576.

54. R. Virchow, "Uber das granulierte Ansehen der Wanderungen der Gehirnsventrikel," *Allgemeine Zeitschrift für Psychiatrie*, vol. 3 (1846): 242–250.

55. Hydén (1959), op. cit.

56. W. J. J. Nauta and M. Feirtag, "The organisation of the brain," *Scientific American* (September 1979): 88–111.

57. H. Hillman, *Cellular Structure of the Mammalian Nervous System*, Lancaster: MTP Press (1986).

58. H. Hillman, "The anatomical synapse by light and electron microscopy," *Medical Hypotheses*, vol. 17 (1985): 1–32.

59. B. Katz, *The Release of Neural Transmitter Substances* (Sherrington Lecture), Liverpool: Liverpool University Press (1969).

60. H. Hillman, "A re-examination of the vesicle hypothesis of transmission in relation to its applicability to the mammalian central nervous system," *Physiological Chemistry and Physics and Medical NMR*, vol. 23 (1991): 177–198.

61. H. Hillman, "A new hypothesis for electrical transmission in the mammalian central nervous system," *Medical Hypotheses*, vol. 34 (1991): 220–224.

62. Senate, University of Surrey, (8/87) (30 September 1987): 1–20.

63. Hillman and Jarman, op. cit.

64. H. Hillman, *The Case for New Paradigms in Cell Biology and in Neurobiology*, Lampeter: Mellen Press (1991).

65. H. Hillman, "Resistance to the spread of unpopular academic findings and views in liberal societies, including a personal case," *Accountability in Research*, vol. 1 (1991): 259–272.

66. H. Hillman, "How to make medical research more successful," *The Practitioner*, vol. 231 (1987): 998–1003; 1403–1409.

67. H. Hillman, "The modern tendency to conformity with proposals for reversing it," *New Humanist*, vol. 109 (1993): 13–15.

68. H. Hillman, "Honest research," *Science and Engineering Ethics*, vol. 1 (1995).

69. See notes 24, 25, 38, 52, 57, 58, 60 and 64.

70. Hillman and Sartory (1977), *Perception*, op. cit.; Hillman and Sartory (1980), *The Living Cell*, op. cit.; Hillman (1986), *The Cellular Structure of the Mammalian Nervous System*, op. cit.

MICHAEL MALLORY
GORDON MORAN

6

The Guido Riccio Controversy in Art History

INTRODUCTION

The Guido Riccio is a fresco, or wall painting, in the city of Siena in Italy. Guido Riccio is short for the full name, *Guido Riccio da Fogliano at the Siege of Montemassi*. The standard view has long been that Simone Martini, a famous painter from Siena, painted the entire fresco in 1328–1330, a view adopted by generations of scholars and repeated in many textbooks, guidebooks, reference books, and classroom lectures. This was also our own view until 1977, when we proposed that a portion of the fresco, namely the horse and rider, was painted by a close follower of Simone Martini in 1352, while still attributing the rest of the painting to Simone.

On the face of it, it would seem that the Guido Riccio discussion would be an obvious candidate to remain limited to a small group of academics with a specialization in Sienese painting, with discussion involving fine points of an intellectual, if not erudite, nature. After all, a portion of a painting was being dated twenty years later then commonly believed, and this specific portion was being attributed to a close follower of a famous painter, rather than to the famous painter. Yet, for various reasons, the issue became hotly contested, immediately got into the mass media, and swiftly escalated into what several writers have described as "the case of the century," or the "enigma" of the century in art history.

The controversy—or "war," as one scholar put it—over Guido Riccio has raged for more than a decade. Hence, this chapter can be no more than a progress report, revealing some highlights from the history of the controversy. We hope to provide some insights into how experts react to upsetting hypotheses and also into how their various tactics and maneuvers can be challenged both inside and outside academia.

The lengthy duration of the Guido Riccio controversy has given us the time and the opportunity to study other academic controversies of the past and present, and to compare notes with other scholars who are involved in ongoing disputes of their own. Such comparisons have enabled us to detect, even predict, patterns of behavior on the part of experts as they attempt to overcome challenges. These patterns include: the suppression and censorship of the challengers' ideas from scholarly conferences, symposia, and journals; personal attacks, including insults, retaliation, and ostracism, against the challengers; and secrecy, instead of open discussion and debate.

VESTED INTERESTS AND MOTIVATION

As the Guido Riccio controversy has progressed, an increasing number of persons have taken an interest in it, including undergraduate and graduate students, art students, alumni groups, culturally minded tourist groups, and local Sienese civic groups. Frequently we are asked: "Why can't they accept the truth?"; "Why can't they admit Simone didn't paint the famous Guido Riccio fresco?"; "Why wouldn't they let you be on the program of the Simone Martini conference in Siena in 1985?"; "Why don't they want to uncover the fourteenth century frescoes that might be hidden under the plaster on the walls of the same room where the Guido Riccio painting is located?" "They" in these questions are the various persons who have vested interests in confirming the standard view of the Guido Riccio.

The famous fresco is located in the main council room of the museum of the Palazzo Pubblico in Siena, situated on one side of the famous town piazza known as the "Campo." A large wall painting or mural, painted in the fresco technique, it is one of the most—if not *the* most—famous painting in Siena, and in the history of Italian Renaissance art. It is listed in many art history textbooks, reference books, and monographs, as well as guidebooks and brochures, as one of the few documented works of the fourteenth-century Sienese painter Simone Martini. Simone, in turn, is regarded as one of the most famous and important painters of the Late Medieval and Early Renaissance period both in Siena and in Italy.

Reproductions of the famous Guido Riccio—particularly its image of the horse and rider—are found on posters for the local Sienese tourist agency, on the covers of guidebooks and textbooks, postcards, plates, ash trays, cookie box covers, lampshades, bathroom tiles, calenders, wine bottle labels, and even blankets. It seems clear that much of the popularity of the image of Guido Riccio on horseback stems from the belief that it was painted by Simone Martini. In fact, generations of Sienese children have been brought into the Palazzo Pubblico museum in Siena by their school teachers and told that Simone Martini painted the image of Guido Riccio on horseback for the glory of the Sienese Republic of the fourteenth century, or some remark to that effect. The painting has become as part of Sienese pride in its history and artistic patrimony and heritage. Moreover, the local Sienese government controls, as a virtual monopoly, the touristic guided tours of the city and its museums and grants licenses to only a select and limited number of official guides who claim exclusive rights to show groups of tourists around the city. Over the years, these "official guides" have told many thousands—if not millions—of tourists that the famous Guido Riccio fresco was painted by Simone Martini, and some continue to do so. And, to be sure, for many years, numerous art history professors in art history classes in universities around the world have waxed eloquently about the painting, describing it as Simone's masterpiece.

The producers of Chianti Classico and Brunello wines have a vested interest in Guido Riccio since he appears on their wine labels. Even more so the city government officials who are the "owners" of the painting located in their city hall and the officials of the Soprintendenza (the federal Italian government agency with responsibility for preservation of works of art in Italy) who must preserve it have a vested interest in the Guido Riccio. Obviously the local tourist board has a vested interest in it, since the painting is one of the big attractions of Siena. And the citizens of Siena themselves, and their cultural institutions, such as the Accademia degli Intronati, have vested interests based on their pride in their city, its history and its artistic heritage. As the Guido Riccio controversy developed, other persons and institutions became involved, such as the powerful Monte dei Paschi Bank, which financed a monograph book on the Palazzo Pubblico, certain art libraries and art library associations which in turn involved the German government, the editors of scholarly journals and specialized encyclopedias, and certain academic professional societies, such as the College Art Association of America, a member of the American Council of Learned Societies.

Rather than trying to guess what prompted the actions, reactions, and maneuvers of these persons and organizations, we will describe and analyze events and situations that have actually taken place. Readers can then decide for themselves what the motivation might have been.

HISTORICAL AND ART HISTORICAL BACKGROUND

Siena is a small city in central Italy, about an hour's drive south of the larger and more well-known city of Florence (Firenze), its historic rival. Siena's history most likely extends to Etruscan times, and it is referred to in Roman literature and historical writings. In Medieval times, Siena developed into something of a political and economic power in its own right, in part through international banking. In fact, in the fourteenth century Siena was a flourishing city-state, expanding its territory in all directions, to the west as far as the Mediterranean coast, to the north to the Chianti region, to the south as far as Mt. Amiata, and to the east into the Val di Chiana. This territorial expansion revived centuries' old conflicts with the leaders of the feudal lords who owned the castles and territory in the areas of the Sienese countryside. During the twelfth, thirteenth and fourteenth centuries military conquests or financial transactions by the Sienese government brought many castles and their surrounding land into its jurisdiction and under its military control.

It is precisely during this territorial expansion program that the Guido Riccio story begins. Early in the fourteenth century, the Sienese government had decided to depict in wall paintings castles that the government had recently taken under its jurisdiction that were deemed to be important and strategic from a military/political standpoint. These castle depictions eventually covered large portions of two walls of the main council room of the Palazzo Pubblico. It is documented that at least seven castles were painted on the walls as part of this artistic-political propaganda program. The first castles were painted before 1314 and they continued to be painted at least through 1331. Simone Martini painted at least four, namely Montemassi and Sassoforte in 1330 and Arcidosso and Castel del Piano in 1331. Secondary sources state that other castles, including Ansedonia and Sinalunga, were also painted as part of this series.

We feel that as many as twenty castles were painted in the room and, moreover, we believe that several—if not many—of these paintings are still preserved, hidden under the plaster of later paintings that currently decorate the room.

Until 1980, it was believed that not only was the famous *Guido Riccio at the Siege of Montemassi* one of the castle depictions by Simone Martini, but that it was the only one of the series to have survived. Specifically, it was associated with the document stating that Simone painted the castle of Montemassi in 1330. What was at best a tentative hypothesis soon was presented as fact in textbooks and guidebooks and in classroom lectures. Guido Riccio became entrenched in the history of art as a documented masterpiece.

In 1980, a second fresco was uncovered from under modern plaster on the same wall as the Guido Riccio. This newly-uncovered painting has subsequently been regarded unanimously by all scholars who have written on the subject as being one of the castles of the castle program described above. Its discovery raised hopes that other masterpieces of fourteenth century Sienese painting might still be hidden under the plaster in the room, waiting to be uncovered for the world to see. It was also soon evident that the Guido Riccio was in every respect very different from the newly discovered work that everyone regarded as original. Doubts about the origin of the famous work began to rise.

CONNOISSEURSHIP IN ART HISTORY

In the field of art history, expert claims for intellectual and professional superiority often rest on their skills as connoisseurs. They profess to be able to "see" and to attribute works of art better than others, especially the press and the public. Perhaps the classic example of this supposed higher sensibility is illustrated by a tale involving one of the greatest acknowledged connoisseurs of Italian paintings, Bernard Berenson. It is said that a person had a painting thought to be by the famous fourteenth century painter Duccio, an earlier contemporary of Simone Martini, and brought the painting to Berenson for his opinion. Berenson, according to the story, said the painting was *not* by Duccio. When the owner of the painter asked Berenson how he could be sure of his opinion, Berenson allegedly replied that had it been by Duccio he would have swooned with aesthetic rapture.

This type of reply intimidates the nonexpert. If a group of "experts"—self-appointed or otherwise—band together and agree on attributions, it is virtually impossible to mount a challenge. When "experts" disagree with each other, however, the subjective nature of connoisseurship becomes obvious. Different attributions for the same work make it clear that one or more of the experts is mistaken. And once an expert is shown to be fallible, doubts begin to arise about connoisseurship being *the* definitive methodology for art history.

An analysis of how connoisseurship has been applied to the Guido Riccio case was made by Joseph Falcone: "In reading all of the material of the scholarly debate between Mr. Moran and what I call the 'normative art history community,' . . . I have come to some interesting conclusions about how art historians debate, including their language and acceptable criteria for evidence in support of their arguments. . . . The use of specific evidence is important to note in the Guidoriccio debate, as it seems to be representative of a normal research tradition. . . . The most important form of evidence that is considered appropriate to use in debate in the discipline of art history is stylistic evidence. . . . 'Normal' art historians were bounded in the Guidoriccio case, as we shall see, by this shared criterion for research. . . . Because Mr. Moran throughout the intellectual debate has advocated the use of technical-empirical and historical evidence as the basis for his arguments, he is operating outside of the 'normal' tradition of art history."[1]

The earlier-mentioned discovery in 1980 of another very different fresco on the same wall as the Guido Riccio caused havoc among the experts. The juxtaposition of the frescoes, with the Guido Riccio work overlapping the one below it, seemed to demonstrate unequivocally that the Guido Riccio was not part of the castle series and was of a more recent origin. When several experts involved in the debate were asked to state how they viewed the chronological relationships of the two works, they refused to answer. Connoisseurship may have backfired on the experts this time, for it seemed clear that many non-experts—such as students, the press, and the public—"saw" what the experts claim they could not see, or didn't want to admit that they saw.

PRELUDES TO THE CHALLENGE

When, during the late nineteenth and early twentieth century, art history became a bona fide scholarly discipline, the Guido Riccio fresco had already become accepted as a documented masterpiece by Simone Martini by guidebooks published in Siena and elsewhere. Despite appeals to the authority of connoisseurship and stylistic evidence that have since been heard in the Guido Riccio debate, it appears that the Simone Martini attribution for the famous fresco slipped into the art history literature without academics and connoisseurs making close stylistic analyses to support such an attribution.

In fact, some doubts about the attribution did surface. Early in this century, a widely-recognized expert of Italian painting, Adolfo Venturi, wrote in his monumental work *Storia dell'Arte Italiana* that he did not believe that the figure of Guido Riccio on horseback was painted by Simone Martini, but rather that it was painted after Simone's death as a symbolic figure of power above a revolving map that was placed on the same wall. Venturi's doubts, however, appeared in a footnote and were not cited.

Furthermore, in an Early Italian Renaissance Art course taught in 1957–1958, Professor Hellmut Wohl expressed doubts about the Guido Riccio when he discussed Simone Martini, though he did not categorically exclude it as a work by Simone. We were both students in Wohl's course and it may well be that he planted the seeds of doubt for the sustained challenge which we mounted a couple of decades later. Who knows how many other scholars had similar questions about the painting, but did not follow up on them in a way that allowed discussion within the scholarly communication system in art history?

THE CHALLENGE BEGINS (BY CHANCE!)

The Guido Riccio challenge originated by chance, almost as an afterthought. We are both specialists in Sienese painting, particularly of the fourteenth century, and for a long time we went along with the traditional attribution. In 1976–1977, one of us (Moran) undertook a study of the overlapping of attributions to Luca di Tommé and Niccolò di Ser Sozzo of various paintings dating from the second half of the fourteenth century. Originally, the Guido Riccio did not enter in any way whatsoever in this study.

During the course of the Luca di Tommé and Niccolò di Ser Sozzo research, Moran found a document in the Archivio di Stato di Siena referring to a payment, allegedly of 1346–1347, to "Simone dipegnitore," for a painting on the Porta Camollia in Siena. This document aroused great curiosity, inasmuch as Simone Martini had died in 1344. There had been considerable discussion in art history literature about whether Simone Martini painted the Porta Romana or the Porta Camollia. With the discovery of this document, Moran hypothesized that perhaps there had been another painter, a close contemporary of Simone Martini, also named Simone, and that this coincidence might have caused some of the confusion about who painted Porta Camollia. Since Moran believed that this unpublished document might be an important key to resolving this art historical problem, he undertook further investigations in this direction, putting the Luca di Tommé and Niccolò di Ser Sozzo project on the back burner indefinitely. These further investigations regarding "Simone dipegnitore" eventually involved the famous Guido Riccio painting, and a series of incongruities and anomalies in this work came to light. Ironically, it turns out that there might have been an error in the compilation of the archival records and that the "Simone dipegnitore" mentioned in the aforementioned document might actually refer to Simone Martini after all. Be that as it may, the challenge to Guido Riccio was about to begin.

CONTENT AND FORM OF THE ORIGINAL CHALLENGE

In 1977, the challenge did not include the entire painting, only the horse and rider. We hypothesized that the equestrian figure of Guido Riccio was a posthumous memorial portrait painted in 1352 or soon thereafter superimposed on the documented depiction of the castle of Montemassi.

Documentary and iconographical considerations were already becoming complex in the first year of the challenge, but at that time we focussed our attention on two main observations:

1. From the standpoint of space and setting, the horse and rider did not seem to be integrated into the scene of the siege of Montemassi. Instead of being part of the narrative, it looks as if the horse and rider are floating across the front of the picture plane, with the left front hoof of the horse resting on the border of the fresco at a point where the stakes of a wooden fence recede below the border, and with the other three legs of the horse suspended in mid-air above a valley which swoops below.
2. Guido Riccio left Siena in disgrace in 1333. According to documents and secondary sources, he let the enemy escape when he had them within striking distance, he was bribed by the enemy,

he let supplies get in when he had them besieged at Arcidosso in 1331, and he was accused of cowardice. Also, he left Siena with considerable unpaid debts. Based on these notices, we thought that if Simone Martini had in 1330 painted a portrait of Guido Riccio, a mercenary soldier who sold his services, this portrait would have been painted over or destroyed in 1333 or soon thereafter. (We later discovered that from around 1333 to around 1350, Guido Riccio was a leader of forces that were enemies of Siena, a fact which reinforced the theory that the Sienese government would have erased any portrait of him in their main council room after 1333.)

The Sienese hired Guido Riccio again in 1351 as their military chief, perhaps thereby buying off a threat to their security. He died in office several months later and was given an elaborate military funeral by the Sienese government, for which the services of some Sienese painters were engaged. We proposed that the portrait we see in Siena today was painted in the context of the funeral celebration as a memorial portrait of a soldier of fortune who had died while in service for Siena. But since Simone Martini himself had died in 1344, the artist of the portrait must have been someone else, perhaps a close follower of Simone. (We now prefer Venturi's theory to our own, as does Professor Federico Zeri.)

It did not take long for the first reactions to take place.

THE FIRST ATTEMPTS BY THE EXPERTS
TO SQUASH THE HERETICAL THEORY

One day in Siena in 1977, Moran ran into Alberto Cornice, an official of the Soprintendenza, and told him of the new theory about Guido Riccio. At the advice of Professor Ulrich Middledorf, the former Director of the Kunsthistorisches Institut in Florence, he made an informal request to have the opportunity to study the fresco closely. Eventually, a portable scaffold was put up in front of the fresco and the Soprintenza requested the restorer Donato Martelli to make some preliminary examinations. Several scholars (experts) were also invited to view the fresco from the scaffold.

During these first investigations, Martelli removed some of the plaster on the wall just below the Guido Riccio and found traces of another fresco which extended underneath the famous fresco's lower border. Since the Guido Riccio fresco overlapped the newly discovered one, the former was later in date than the latter. When the invited scholars called up to Martelli to ask if he had found anything of importance, the restorer, with a sense of humor that might have masked his own suspicions about Guido Riccio's portrait, replied that he had not found anything unusual, merely a portrait of Garibaldi, the nineteenth century Italian military and political hero, underneath the famous fresco.

A few days later, the office of the Soprintendenza, headed by Piero Torriti, called in the famous restorer Leonetto Tintori to make further investigations. Meanwhile, Cornice telephoned to say that a friend of his, Serafina Baglioni, a journalist in Siena, was interested in knowing more about the new Guido Riccio theory. An interview was arranged in front of the fresco on the day Tintori and his colleague were carrying out their investigations.

Baglioni's article, entitled "Simone Martini Contested," appeared in the 4 October 1977 edition of the nationally distributed *La Nazione*.[2] The experts and the political authorities in Siena were taken by surprise by the sudden "onore della cronaca" (honor by means of press coverage) for such a heretical theory. From this point on, the press, particularly the Sienese press, has played an active role in the controversy, with well over one hundred articles on the subject. For the most part, the press has shown what we consider to be a great sense of responsibility throughout the controversy, and the Sienese journalists have shown an extraordinary openness for continual up-to-date coverage of our findings and ideas.

The following day, the Florentine journalist Wanda Lattes' article, entitled "The Mystery of Simone Martini," appeared in *La Nazione*.[3] She reported the opinions of various experts and acknowledged the existence of real problems of art criticism, attribution, and dating, and conservation regarding the painting. She detected two trends among the replies she elicited: one group expected an academic debate to resolve the question, while the other group was determined to defend the traditional attribution. Among the latter, Lattes quoted Professor Giovanni Previtali at length, and he soon became the unquestioned leader of the experts who dedicated themselves to defending the attribution to Simone regardless of the evidence.

The first organized attempt to squash the heretical theory about Guido Riccio occurred at a round table discussion held in Siena in early November 1977 at the Accademia degli Intronati. Invited speakers included Moran and a number of experts of Italian art. This event was open to the public and many Sienese citizens who were merely curious or even shocked or outraged about the new theory also showed up. It was an overflow crowd, in some ways reminiscent of a partisan hometown crowd for a sporting event.

Although promoted as a "round table" session to discuss the issue, at least two Sienese have told us that the forum was organized with the intention of refuting the heretical theory, to nip it in the bud, as it were, in a civilized and scholarly manner. And with Professor Enzo Carli of Siena acting as chairman of the event, there was every assurance that the discussion would be carried out with open, civilized exchange among the participants. But as hard as they tried, the experts could not dispel the doubts. Most difficult to rebut was the idea that had Simone Martini painted a portrait of Guido Riccio in 1330, this portrait would have been erased after 1333 when Guido Riccio fell into disgrace. Once again, *La Nazione* covered the event with a long article entitled "Enigmatic Guido Riccio." The last paragraph begins: "The debate is not closed."[4]

SUBSEQUENT REACTIONS, INCLUDING PERSONAL INSULTS

Not all of the experts and other persons involved have abided by the same high professional standards that Professor Carli maintained at the round table discussion and subsequently throughout the controversy. Although Professor Previtali made some bitter comments, the round table chaired by Carli provided no indication of the future reactions that were to take place as the debate escalated and intensified.

Insults of various types against one or both of us became an inherent part of the Guido Riccio controversy. Here are some selected examples:

"Go back to America, by boat . . . and to the next presumptuous person who comes here we'll tell him to his pig's snout that we don't want to give him a little glory . . ."—Arrigo Pecchioli.[5]

". . . the absurd and defamatory accusations . . . published by Gordon Moran and Michael Mallory . . ."—Piero Torriti.[6]

". . . Gordon . . . gets more pushy, more paranoid, more upset . . . I think if he were a genuine art historian . . . he would do like I would do. . . . He just thinks all the time about this case. . . . He's become almost insanely obsessed with it. I think he has taken it over the brink."—Professor Samuel Edgerton.[7]

". . . It is only the invention of a non-expert who has not found anyone who agrees with him. This American was mistaken from the beginning . . . then he had to eat his words, and he is spending all his life trying to demonstrate that the fresco is not by Simone. Poor man, by now he has taken on the form of a mono-maniac."—Professor Giovanni Previtali.[8]

". . . Professor Bellosi, in an interview, declares that perhaps Moran has become insane . . ."—Giorgio Sacchi.[9]

". . . the two 'monomaniacs' (I'm speaking of Gordon Moran and Michael Mallory) . . ."—Arrigo Pecchioli.[10]

"Moran published his theories about the 'Guido Riccio' in 1977. . . . From the start, he was denounced by the Italian art establishment in vicious terms; he was called a CIA agent, a monomaniac and a paranoid."—Jacob Young with Lin Widmann.[11]

". . . the tenacious attempts to contest the hypotheses of Moran . . . immediately were transferred to a personal level, with bizarre objections to the effect that Moran was not a real historian, but a lightweight dilettante on the subject matter . . ."—Marco Carminati.[12]

"Professors at the University of Siena dismissed him with thinly veiled condescension. . . . Zeri . . . got right to the point. . . . 'Moran is not a Sienese, he's not a member of this inbred confraternity of scholars and he's an American. Therefore, the Sienese professors feel that he has no right to his opinions.'"—Jane Boutwell.[13]

During the course of our research into the Guido Riccio problem, we have formulated no less than *sixty* reasons—historical, documentary, technical-scientific, stylistic, and iconographical—to doubt the traditional attribution. Despite our findings, the above list of comments indicates that some experts involved in the controversy have chosen to treat us not as scholars involved in critical inquiry and discussion concerning a subject of mutual interest but, rather, to treat us as unqualified, crazy outsiders who are trying to intrude upon their writings on and teaching of art history. As might be expected in such a case, some of the experts involved tried to censor our ideas from scholarly journals in the field.

SUPPRESSION AND CENSORSHIP

The rhetoric of academia treats academic freedom as a sacred cow. If material that has been censored becomes known, however, all the more attention is drawn to it.

In the Guido Riccio controversy, there have been well-documented efforts to censor our ideas. Our writings have been rejected by some leading art history journals and we have been excluded as speakers during Guido Riccio discussions. In the longer run, attempts to censor our ideas have failed miserably. If our ideas were rejected by a scholarly journal, they would appear somewhere else. And when we were excluded from the program of a scholarly conference, our absence was noted by the press and our ideas gained even wider coverage.

Leading art history journals in England (*Burlington Magazine*), Germany (*Zeitschrift für Kunstgeschichte*), Italy (*Rivista d'Arte*), and America (*Art Bulletin*) rejected our articles on Guido Riccio. The editors did not question our evidence let alone try to refute it. Instead, they objected to the length (even though they recently published articles longer and shorter than ours), format, and style.

The editor of *Burlington Magazine* sent our paper to a referee for her opinion about whether our article should be published. The issue of Guido Riccio was so important, she is reported to have said, that the new information in our article should be accepted. In spite of this, it was rejected. The editor of *Rivista d'Arte* rejected our article on the basis that it should appear as a letter to the editor. When we rewrote the piece in the form of a letter to the editor as he had suggested, he quickly rejected it, claiming that his journal does not publish letters to the editor. And so it went.

The refusal to allow new hypotheses and new findings to be presented and discussed during scholarly conferences is another form of academic censorship. A clamorous case occurred in 1985 at a conference held in Siena to study the works of Simone Martini. The conference was jointly organized by the University of Siena, the local government of Siena, and the office of the Soprintendenza of Siena. Hearing that scholars were being requested to participate on the program of the conference, we asked permission to be on the speakers' program to present new evidence that we had recently discovered but not yet published. The mayor of Siena, Mazzone della Stella, replied that the Organizing Committee had rejected our request. We then wrote to several members of the Organizing Committee, asking them if they had personally made a negative judgement against us in the Organizing Committee's rejection. Some of the members replied that not only had they not voted against us, but that they were not even aware that other members of the committee had rejected us. One member of the committee, Professor M. Frinta, wrote to us that he considered our rejection in this case to be an example of "foul play."

We then enquired precisely *why* our request had been denied. After all, among works attributed to Simone Martini, the Guido

Riccio was currently the focus of considerable attention among art historians and we felt our new findings were significant and timely. The reason for our rejection was explained by Professor Bellosi, one of the Inner Committee of the Organizing Committee that disallowed our request to speak, to another member of the Organizing Committee, Professor Miklos Boskovits, who conveyed it to us. According to Boskovits, Bellosi stated that by then scholars knew where each side stood on the issue of Guido Riccio, that the subject had been worked over in detail recently, and that there should be a pause for reflection. Boskovits said he agreed with the reasoning behind the decision to exclude us from the program.

However, after having kept us off the program because Guido Riccio was not to be discussed, the Inner Committee included one of their own members, Soprintendente Piero Torriti, on the program to give a long talk on the Guido Riccio situation in which he attempted to refute our views. Once revealed, the hypocrisy of this particular instance of censorship was evident to many in Siena, including the media and the art historical community.

A more subtle form of de facto suppression existed—intentionally or unwittingly—by making certain materials far less accessible in at least one library. At the Kunsthistorisches Institut in Florence, Italy, where Professor Max Seidel and Irene Heueck, two members of the Organizing Committee's Inner Committee cited above, hold powerful positions, all publications challenging the traditional interpretation of the Guido Riccio remained unindexed for a long period of time. Later, after Moran exposed this situation at a conference of art librarians, the Director of the Kunsthistorisches Institute tried to take retaliatory action by not renewing his library card and making him sign a declaration agreeing to keep silent in order to have it renewed.

OVERCOMING CENSORSHIP AND SUPPRESSION

Since academic censorship is contrary to the tenets of academic freedom, the mere suggestion that ideas have been censored might put the censors and suppressors on the defensive, while widespread exposure of academic censorship might put them on the run and might also cause additional potential future would-be censors to refrain from censorship. At the same time, within a given academic discipline there may well be some editors of scholarly journals willing to publish the material that has been censored by other journals, either out of a sense of justice and fair play, out of a desire to end up on the winning side, or out of a true belief in open academic debate. Also, some scholars who have felt the effect of censorship imposed upon them might well increase their efforts to get their views known.

In fact, some individuals went out of their way to offer to publish our views after the debate had intensified in 1980–1981. Among the first was Giorgio Sacchi, a Sienese artist who heads *Notizie d'Arte*. Although this journal appears only sporadically and is chronically short of funds, it has given extensive coverage to the Guido Riccio debate, particularly the issues of August 1981 and September 1985, the latter of which contains an article of ours giving an update on the controversy.[14] Sacchi became outraged by the censorship he perceived to be taking place and he has been a thorn in the side of the official group of experts. In addition, Professor Miles Chappell invited us to publish our views in a journal, *Studies in Iconography*, for which he was Acting Editor.[15]

After we had been excluded from the Speakers' Program of the Simone Martini conference in Siena, and in the wake of the rejection of our article by *Burlington Magazine*, the leadership of the Harvard University Center for Italian Renaissance Studies in Florence, Italy, known as Villa I Tatti, offered to give us some assistance. Professor Craig Smyth, Director at the time, and Eve Borsook, a Research Associate there, helped revise the article that *Burlington* turned down and they strongly recommended that the journal publish it. Thanks to their sense of fair play, our article appeared in the April 1986 issue of *Burlington*.[16]

When *Rivista d'Arte* rejected our article in which we attempted to correct Professor Seidel's incomplete transcriptions of a document regarding the 1314 submission of Giuncarico to Siena, we were invited to submit the same material to *La Gazzetta di Siena*, a local newspaper, which published it in 1983.[17] Eventually we managed to get some of this information in a footnote of our *Burlington Magazine* article when correcting the proofs. Thus, serious flaws in Professor Seidel's research on Guido Riccio were finally revealed in the art history literature, but only partially and in a footnote and only well after these same defects had been fully exposed in a local newspaper for all Siena citizens to see.

In Siena, we requested permission to write a short rebuttal article in the scholarly publication of the Accademia Senese Degli Intronati, *Bulletino Senese di Storia Patria*, and our article was published.[18] In this case, the editorial board included Professors

Enzo Carli, Mario Ascheri, and Giuliano Catoni, all of whom disapproved of the censorious treatment that we had been receiving. In fact, it turned out that Professor Ascheri was on a receiving end of censorship himself in the Guido Riccio case. During the public discussion period at the Simone Martini conference in Siena, he pointed out that it was highly unlikely, if not impossible, for the large equestrian portrait of Guido Riccio to have been painted during Simone Martini's lifetime. As a leading expert in the field of the history of Medieval law, Ascheri's arguments were based on the political and legal realities of that time. When it became apparent that his ideas supported our hypotheses and cast very strong doubts on the official view, he was interrupted by art historian Ferdinando Bologna and told, in front of the large audience, that he could not talk like that at an art history conference. Ascheri subsequently published several articles in Sienese newspapers and magazines developing his ideas, and they were republished in a recent book.[19]

The censorship imposed by the experts against anyone who dared oppose their official views about Guido Riccio at the Simone Martini conference created a negative impression among Sienese citizens, the Sienese press, and the national media. This situation only worsened when, during his talk on Guido Riccio at the conference, Torriti repeatedly castigated the press for their interference in the Guido Riccio question. The press reaction to the experts is perhaps best summed up by the title of an article which appeared in Siena, "Guido Riccio Drowns in a Sea of Intolerance."[20] Soon after, we were invited to present our views at various civic and cultural groups in Siena, including Rotary Club Siena Est and the cultural clubs Hobbit and Ignacio Silone.

The Kunsthistorisches Institut's selective indexing of the Guido Riccio literature was presented as a case study for art librarians at the 2nd European Conference of the Art Libraries of the International Federation of Library Associations in Amsterdam in October 1986. A few years later, the papers given at the conference were published. As mentioned above, Joseph Falcone later wrote about this situation, as did John Swan, Head Librarian at Bennington College, and a leader in Intellectual Freedom Roundtable activities of the American Library Association.[21]

It seems to us, then, that attempts at censorship and suppression in the case of Guido Riccio have turned out to be counterproductive and embarrassing for the experts. This does not mean, however, that they will not try again.

FALSIFICATIONS AND STONEWALLING

It is our opinion that some of the research put forth by the experts in most Guido Riccio studies contains serious falsifications of evidence. Whether these misleading errors were made intentionally or unwittingly is not for us to say. It does appear that our attempts to describe and expose falsifications that we have detected results in our censorship by those in authority. An attempt to illustrate *all* of the falsifications we have detected so far would take up more space than this chapter allows, so we will give one example to indicate just how preposterous some of the scholarship of the experts has become and how difficult it has been to make this known to art historians.

In the March 1987 issue of *Burlington Magazine*, we pointed out in a letter to the editor that a portion of the lower border of the Guido Riccio fresco had been destroyed during the 1980–1981 restoration and that this destroyed portion of the fresco constituted crucial evidence that could no longer be studied by scholars.[22] Piero Torriti wrote in reply that we had made "absurd and defamatory accusations" and in his letter he purported to "refute" them "once and for all."[23] He included a color photograph as part of his letter and claimed that the portion of the fresco in question still exists in its original form. But anyone who looks at either the fresco or the photograph that Torriti published can easily see for themselves that the portion of the border in question has disappeared. As in the fable *The Emperor's New Suit of Clothes* by Hans Christian Anderson, art historians are asked to "see" something that has been removed.

We regarded Torriti's charges against us as completely false and we attempted to publish a rebuttal in which we pointed out the obvious. We were met with adamant and persistent stonewalling by *Burlington* editor Caroline Elam, who wrote no less than five rejection letters in her efforts to keep our response out of the journal. We countered with a series of letters and open letters to a widening group of academics interested in problems of peer review, scholarly communication, and academic ethics. Eventually the Board of Directors of *Burlington Magazine* reversed Elam's decision and we were allowed to publish our reply.[24]

In the *Burlington* case, stonewalling was not confined to the editor. Among other officials of the journal, we wrote to Sir Brinsley Ford, a Trustee of The Burlington Magazine Foundation, enquiring if he thought that scholars who had been charged with having made "absurd and defamatory accusations" should be allowed to reply to the charges. He quickly responded: "You have made accusations to which Professor Torriti had the right to reply, and that, in my opinion, should be the end of the matter so far as the Burlington is concerned." Just what, in his opinion, did our accusations consist of and against whom were they made, we replied. Despite a follow up letter, we never heard from him again. Eventually Elam intervened, requesting that we not "badger" Brinsley Ford further because he was about eighty years old and did not have secretarial help. So here we had a situation in which a *Burlington* Trustee was in no way prevented by age or lack of secretarial help from firing off an immediate reply that leveled false charges against us, but when he was requested to back up his charges, suddenly age and lack of secretarial help prevented him from doing so.

In his recent monograph on Simone Martini, Professor Andrew Martindale wrote that he believes the famous Guido Riccio fresco and the fresco discovered in 1980–1981 were both commissioned to Simone Martini in a time span of about eighteen months between 1331 and 1333.[25] At one point, believing that the reviewers and supporters of Martindale's views were perpetuating an obvious error in the scholarly literature, we wrote specific questions to some of Martindale's reviewers and supporters. We received few replies. We feel that this sort of stonewalling is very revealing about what is currently going on concerning a crucial aspect of the Guido Riccio story.

HOW AND WHEN WILL THE CONTROVERSY END?

Debate and discussion about the two frescoes have been raging for more than a decade without resolution. Hundreds of writings, including newspaper articles, letters to the editor, articles in scholarly journals, and articles in popular magazines have been devoted to the subject. It is very unusual in art history that a question of attribution and dating for two works of art should occupy so much time and space in the scholarly literature and in the mass media.

We should recall at the same time that the Guido Riccio is a kind of secular icon, the delight of museum-going tourists and student groups who discuss it and read about it every year. Also, there has been the desire among many Sienese citizens to be kept up to date on any new developments in the controversy. Moreover, the Guido Riccio case has expanded beyond a narrow art historical issue of dating and attribution to enter the arena of academic ethics, peer review, scholarly communication, and the sociology of higher education. Therefore, it seems unlikely that the controversy will die down in the near future.

More likely, it will intensify. If so, two consequences seem likely. On the one hand, the multiple attributions for the fresco discovered in 1980 might well lead to confusion among scholars and even to distrust on the part of younger scholars of the traditional methodology of connoisseurship relied upon so heavily by the experts. On the other hand, there might well be a widening gap between a growing number of students and members of the public who no longer accept the Simone Martini attribution and the experts such as Bellosi, Seidel, Polzer, Christiansen, Martindale, Liedtke, Strehlke, and Torriti who seem further and further committed to defending it.

Some indications of these latter developments can already be detected. For example, in a recent monograph on Simone Martini, Cecelia Jannella writes about the fact that various scholars have written different attributions for the fresco discovered in 1980: "This incredible difference of opinions. . . . The observer may be surprised by this variety of attributions, especially since the artists mentioned are all so different. But for the public in general, informed by the unusual amount of space the press devoted to the matter, and also for those who take a professional interest, the main problem was how to form one's own opinion. How disorienting . . . when learned art historians contradict one another so drastically."[26] It seems to us that these contradictions among art historians in this instance, and the disorientation that results among the public, students, and scholars themselves, all serve to prolong and draw attention to the controversy.

The official guidebooks in Siena, the signage in the Museum of the Palazzo Pubblico of Siena, and the audiovisual guide machines in the museum that museum goers use, all maintain the traditional attribution for Guido Riccio, with hardly a word that there has been a long and ongoing controversy. Also, some if not many of the official guides who are licensed by the local government of Siena and who have virtual monopolistic authority to show tourist groups around Siena, spout the official views about Guido Riccio when describing the paintings to the tourists in the Sala del Consiglio of the Palazzo Pubblico. But many of the guides have shown a keen interest in following the controversy closely and somehow the tourists in their groups hear something of the other side of the story and get a pretty good idea of how things stand. Moreover, some guidebooks from outside of Siena, particularly those published in England, report the challenge to the official view. For example, "The Rough Guide" to Tuscany and Umbria relates the following: "The fresco on the opposite wall . . . *Guidoriccio da Fogliano* . . . was until recently also credited to Martini. . . . Art historians, however, have long puzzled over the anachronistic castles—much later in style than the painting's supposed date of 1328—and in the mid-1980s further evidence was found when, during restoration, an earlier fresco was revealed underneath. The current state of the debate is confused, with a number of historians—led by the American Gordon Moran (whom the council for a while banned from the Palazzo Pubblico)—interpreting the *Guidoriccio* as a sixteenth-century fake, the others maintaining that it is a genuine Martini overpainted by subsequent restorers. The newly revealed fresco below the portrait, of two figures in front of a castle, is meanwhile variously attributed to Martini, Duccio and Pietro Lorenzetti."[27] Two standard textbooks, Gardner's *Art Through the Ages* and Hartt's *History of Renaissance Art*, have excluded Guido Riccio as a work by Simone Martini in their most recent editions, with Hartt acknowledging our view that Guido Riccio is a post-sixteenth century work. Also, several other scholars have written their doubts about the traditional Simone Martini attribution, including Zeri, Briganti, Jannella, Frugoni, Redon, Parronchi, and Ascheri.

It appears we are headed for a confrontation between the experts and their official guides on the one hand, and a growing number of students, art historians, Sienese citizens, and museum goers from outside Siena on the other hand. Aleana Altmann, a student from near Geneva, Switzerland, has recently written a term paper on the Guido Riccio controversy. In the course of her studies, she was at a gathering at which some Sienese residents were present. When she approached a man who by appearance seemed quite cultured and asked him what he thought about the Guido Riccio question, he replied to the effect that the officials in Siena were trying hard to maintain and sustain the traditional attribution, but that everyone in Siena knew that the painting is a fake! If this is an accurate assessment, a "showdown" may occur sooner than we ever imagined, with further embarrassment for the experts.[28]

But there is another possibility. If discoveries are made which bring to light other hidden fresco masterpieces from the painted castle cycle in the Palazzo Pubblico, much progress toward resolution of the Guido Riccio question could easily take place. In our view, the combination of these potential additions to art history and the artistic patrimony of Siena as well as to the new chapters in art history that will have to be written might well result in a sort of cultural euphoria that would allow the past bitterness of the controversy to fade away. And in the end everyone would turn out to be on the winning side, including the experts themselves.

TABLE 6.1. Methods of Research and Related Activities

Research for the Guido Riccio painting began with the discovery of a document in the Siena archives (Archivio di State). At that point, the research proceeded in a manner that would be considered normal for art history studies. In addition to a stylistic analysis and study of the painting itself, studies were undertaken to determine what other scholars have observed and written about the painting in the past. Such studies were carried out in art history libraries such as I Tatti (of Harvard University) and the Kunsthistorisches Institute (of the German government), both located in Firenze.

As the studies progressed and as a controversy developed around the painting, the nature of the research involved expanded. Anomalies and anachronisms were detected in various aspects of the painting. As Alice Wohl observed, "But the questions are many, and they are not resolved. . . . The range of issues, involving not merely Trecento painting in Siena but also heraldry, costume, seals, military architecture and the history of warfare, political and social history, topographical illustration, technical expertise, and the interpretations of documents, engages every aspect of the discipline of art history."[29]

Questions relating to genealogy, heraldry, political alliances, and so forth, led to research being conducted in the archives and libraries in varies cities of Italy, including Bologna, Modena, Reggio Emilia, Ferrara, Padova, Milano, and further investigations might include study trips to Verona and Venice.

Other research activity included several on-site inspections and investigations of various castles historically linked, in one way or another, to the Guido Riccio controversy, including Arcidosso, Montemassi, Giuncarico, and Castel del Piano. These investigations revealed what we consider to be gross errors, if not serious misrepresentations, in the published studies of Italo Moretti relating to the topography and orography of Montemassi, and in the studies published by Max Seidel regarding the fortifications of Arcidosso and the topography of Giuncarico.

As the controversy intensified, the mass media, which was involved locally from the start, began to take an interest on an international level, and in addition to submitting our findings and rebuttals for publication in art history journals, we also gave such information during interviews with magazines and newspapers, during television shows and news broadcasts, and also during press conferences that we held recently. *Newsweek* (international edition), *International Herald Tribune, The Economist, The Observer* (London), *The London Times*, RAI (Italian national television), and ABC television are among some of the more widely known members of the mass media which asked to hear our side of the story. Several leading national Italian newspapers also interviewed us, including *La Repubblica, La Stampa, Il Giornale* and *La Nazione*.

In addition, many scholarly groups and universities asked us to give updates on our studies. Mallory gave talks, illustrated with slides, to the annual meetings of the College Art Association of America and of the International Foundation for Art Research as well as to the Institute for Advanced Study at Princeton University, New York University Institute of Fine Arts, Temple University, Wesleyan College, and the like. We gave joint lectures at Harvard University as well as at their I Tatti study center in Firenze. Moran has been requested to speak to various university groups that made field trips to Siena as part of their academic program, including groups from Toronto University, Zurich University, Bokum University, Syracuse University, Pennsylvania State University, Georgetown University, American Institute for Foreign Study, Association of Midwest Colleges, Tulane University, Lewis and Clark College, California State University, and Williams College. Tour groups such as Butterfield and Robinson and the Smithsonian Institute have also shown an interest, as have study groups, including Art History Abroad (London).

Since the range of topics extends beyond fourteenth-century Sienese painting, as Alice Wohl, cited above, mentions, we have come across much information not directly related to the Guido Riccio painting. Some of this information pertains to artists other than Simone Martini. In fact, in the course of our research for the Guido Riccio case we have come across rather startling information relating to studies about the famous sixteenth-century painter Beccafumi and the fifthteenth-century Sienese painters Giovanni di Bindino and Stefano di Giovanni (known as "Sassetta"). We intend to publish more on these subjects in the future, in addition to whatever new significant findings come to light relating to Guido Riccio.

NOTES

1. Joseph Falcone, *Is Knowledge Constituted by Power? The Politics of Knowledge in the Art History Community: A Case Study of 'The Guidoriccio Controversy'*, Honors Thesis in Anthropology, Hamilton College, Clinton, New York (3 May 1991): 3, 7, 8.

2. Serafina Baglioni, "Simone Martini contestato," *La Nazione* (4 October 1977): 4.

3. Wanda Lattes, "Il mistero di Simone Martini," *La Nazione* (5 October 1977): 3.

4. "L'enigmatico Guido Riccio," *La Nazione* (5 November 1977): 4. The quoted sentence was "Il dibattito non é chiuso."

5. Arrigo Pecchioli, "E adesso signor avvocato," *Il Campo* (25 May 1979) (our translation from the Italian).

6. Piero Torriti, *The Burlington Magazine* (July 1989): 485.

7. See Falcone, op. cit.: 49.

8. Eleonora Mariotti, "Da Modì a Guidoriccio," *Nuovo Corriere Senese* (19 September 1984): 6, 7 (our translation from the Italian).

9. Giorgio Sacchi, letter, *Nuovo Corriere Senese* (21 September 1988) (our translation from the Italian). In the interview that Sacchi refers to, published in *Nuovo Corriere Senese* (3 August 1988): 13, it is reported that Bellosi states that it seems to him that Moran has suddenly become insane. Bellosi's words are, ". . . mi sembra che a Gordon Moran abbia dato di volta il cervello."

10. Arrigo Pecchioli, "La Storia Dice: È di Simone," *Il Campo* (29 September 1989): 12 (our translation from the Italian).

11. Jacob Young with Lin Widmann, "Italy's Great Fresco Fracas," *Newsweek* (International edition) (4 February 1985): 49.

12. Marco Carminati, "Un 'affresco un po' frescone," *Il Sole—24 Ore* (31 December 1989): 13 (our translation from the Italian).

13. Jane Boutwell, "Fiasco al Fresco," *Domino* (March 1989): 131–133, 147.

14. Michael Mallory and Gordon Moran, "Aggiornamenti sulla Controversia su Guido Riccio ad il Ciclo di Castelli Dipinti nel Palazzo Pubblico di Siena," *Notizie d'Arte* (September 1985).

15. Michael Mallory and Gordon Moran, "*Guido Riccio da Fogliano*: A Challenge to the Famous Fresco Long Ascribed to Simone Martini and the Discovery of a New One in the Palazzo Pubblico in Siena," *Studies in Iconography*, vol. 7–8 (1981): 1–13; Gordon Moran, "Guido Riccio Da Fogliano: A Controversy Unfolds in the Palazzo Pubblico in Siena," *Studies in Iconography*, vol. 7–8 (1981): 14–20.

16. Michael Mallory and Gordon Moran, "New Evidence Concerning 'Guidoriccio,'" *Burlington Magazine* (April 1986): 250–256.

17. Michael Mallory and Gordon Moran, "Interessante documentazione sul Guido Riccio e sul ciclo di castelli dipinti nel civico palazzo," *La Gazzetta di Siena* (5 November 1983): 4.

18. Michael Mallory and Gordon Moran, "Precisazioni e aggiornamenti sul 'caso' Guido Riccio," *Bulletino Senese di Storia Patria* (1985): 334–343.

19. Mario Ascheri, *Dedicato a Siena*, Siena: Edizioni Il Leccio (1989): 61–80, in the chapter "Arte: Guidoriccio e dintorni."

20. Duccio Balestracci, "Il Guidoriccio annega in mezzo all 'intelleranza,'" *Nuovo Corriere Senese* (3 April 1985).

21. John Swan, "Ethics Inside and Out: The Case of *Guidoriccio*," *Library Trends* (fall, 1991): 258–274. Among his research sources, Swan uses Alice Sedgwick Wohl, "In Siena, an Old Masterpiece Challenged, a New One Discovered," *News from RILA* (February 1984), an important bibliographic survey that helped make our views known in the midst of efforts to suppress them in various scholarly forums and settings.

22. Michael Mallory and Gordon Moran, letter, *Burlington Magazine* (March 1987): 187.

23. Piero Torriti, letter, *Burlington Magazine* (July 1989): 485.

24. Michael Mallory and Gordon Moran, letter, *Burlington Magazine* (January 1991): 37.

25. Andrew Martindale, *Simone Martini*, Oxford: Phaiden Press (1988).

26. Cecelia Jannella, *Simone Martini*, Firenze: Scala (1989): 63.

27. Jonathan Buckley, Tim Jepson and Mark Ellingham, *Tuscany and Umbria (The Rough Guide)*, London, Harrap Columbus (1991): 282.

28. Our latest publications on the subject of the Guido Riccio controversy include: "The Guido Riccio Controversy and Resistance to Critical Thinking," *Syracuse Scholar* (Spring 1991); and "Did Siena Get Its *Carta* Before Its Horse?," *The Journal of Art* (May 1991): 76. In these articles we describe some of the tactics of the experts that are discussed in this chapter.

29. Wohl, op. cit.: 11.

7

Confronting the
Nuclear Power Structure in India

In 1974, after my return from the United States, I joined the most prestigious national university in New Delhi, named after India's first Prime Minister: Jawaharlal Nehru University (JNU). I joined as associate professor with a promise to be elevated in due course to full professor, and was also made chairperson of the recently established Centre for Studies in Science Policy at the School of Social Sciences, JNU.

Those were the years when India was ruled by dynastic Prime Minister Mrs. Indira Gandhi. She was autocratic and corrupt and a shrewd manipulator of money forces and industrial and government machinery to her personal advantage. She wanted India's major scientific and academic institutions and governmental agencies to work towards her dynastic aspirations and, therefore, the new university established in the early 1970s on the U.S. academic pattern was named after her illustrious father, Jawaharlal Nehru. For obtaining research grants or for advancement of your institution, you were supposed to praise the dynasty. It was, therefore, no coincidence that almost all top scientists in India panegyrized the Nehru family.

Within a few months of my joining JNU, Mrs. Gandhi was found guilty of corrupt practices by the Supreme Court. Instead of resigning, she imposed what was called the Emergency. Thousands of citizens and opposition leaders were imprisoned, democratic freedoms were removed, press censorship was imposed, and free associations, meetings, discussions, demonstrations, and unions were banned. It was in that historic moment of national crisis that I decided to dissent in academic circles, and offered a critical voice against corruption and misuse of science for narrow political ends. I challenged the system in academic councils, scientific meetings, seminars, and through my writings. I organized street marches against nuclear power when only a few knew about radiation hazards. I took up the task of educating members of parliament, petitioned the heads of governments against nuclear weapons, and actively opposed India's secret nuclear program.

Though I challenged the political power brokers and their operative influence in scientific and technological decision making, I did not join any political party. Nor did I establish my own political group. All through my years of struggle in India (and earlier in the United States) I sought no political advantage, and basically I remain anti-establishment. Fundamentally, I followed my academic discipline and tried to implement my findings in sociopolitical policy decisions.

Perhaps that was my mistake or perhaps it was my strength. I have no way to measure my success. What mistakes, tactical or otherwise, I made are for the reader to judge. I shall here attempt to narrate my story of confronting the combined forces of political corruption and secret scientific subgovernment of atomic energy in India—including the retribution I received.

THE POLITICS OF NUCLEAR POWER

In 1975–1976, I began studying the sociology of science purely as an academic undertaking. Unaware of the seriousness of problems related to science policy in general and atomic energy policy in particular, I, as chairperson of the Centre for Studies in Science Policy at JNU, organized seminar lectures on energy policy. Among those invited to give a series of lectures was Professor B. D. Nag Chaudhuri, a brilliant nuclear scientist and former director of Saha Institute of Nuclear Physics. Nag Chaudhuri had been the Scientific Adviser to the Prime Minister and headed the Defence Research Organisations. I was to learn later that he was one of the six advisers of Mrs. Gandhi when she decided to explode a nuclear bomb at Pokharan in the Rajasthan desert on 18 May 1974. But now he was my Vice-Chancellor and had shown keen interest in my research. In fact, he was instrumental in appointing me to the chair of Science Policy Studies.

During his lectures on energy policy, Nag Chaudhuri was evasive on issues of nuclear energy. But he encouraged me to investigate and suggested that little work had been done in India on this critically important subject. There existed no critical writings on the subject and most academics, politicians, and the media were not aware of critical assessments of nuclear technology in the west. A powerful and insular group controlled the nuclear establishment, comfortably protected by the Atomic Energy Act 1962, which provided them with all-pervasive legal authority to refuse the public access to any information. The Act forbade any discloure of information which relates to "an existing or proposed plant used or proposed to be used for the purpose of producing, developing or using atomic energy." The Act further read that "No person shall disclose, or obtain or attempt to obtain any information" about nuclear energy activities which was thus restricted under the Act.

Because of such repressive provisions and in view of the strategic importance of the program, no one in my country had ventured to look into the affairs of the nuclear energy department. The patriotic and populist political culture backed by the dynastic regime of Mrs. Gandhi had reinforced denial of public access to critical scientific information.

During the internal Emergency imposed by Mrs. Gandhi from 1975 to 1977, I joined underground activities.[1] During those dark days of political repression, underground activism brought me closer to some of the political bigwigs who later became ministers in the Janata (People's) Government during a brief spell of 1977–1979. In 1978, I came in close contact with Dr. Atma Ram, once the President of Indian National Science Academy and former Director-General of the Council of Scientific and Industrial Research. In the Janata Government, Atma Ram was the Scientific Advisor to Prime Minister Moraraji Desai. He invited me to examine the nuclear establishment critically and report my findings to the new government. Prime Minister Desai and Atma Ram both were known to be Gandhian and antinuclear in their philosophical inclinations.

During 1978–1979, a few months were available to me to peep into the secret chambers of nuclear sub-government. But soon after, due to Mrs. Gandhi's machinations, the Janata government fell, and I received a curt note from a Joint Secretary of the government debarring my visits to nuclear facilities. I was also asked to seek clearance before I made public any information gathered during my visits to nuclear establishment. It was then, when I challenged the order, that I was shown the provisions of the Atomic Energy Act 1962 which bar any disclosure about the nuclear program. Around that time, the U.S. Energy Research and Development Administration downgraded its forecast of twelve hundred nuclear plants of 1,000 MWe capacity each to about four hundred plants by the year 2000, and Sweden adopted a new policy of phasing out its twelve reactors by 2010.

I prepared a comprehensive critical analysis of nuclear power, including a cost-benefit analysis. In a two-part article published by *The Times of India*, I challenged, for the first time in India, the official claims to "clean, safe and cheap source of energy."[2] In the conclusion I stated that the arguments against nuclear technology were too well-established to be rejected as "antiscience." Little debate was permitted by the government on the question of advantages and disadvantages and undue publicity and the glamour attached to the "big bang" grossly distorted our national perception of nuclear reality. Otherwise, "every million earmarked now for the nuclear programme will simply drag us into a quagmire of many more millions within a few years. It is imperative that we consider the economic, industrial and ecological implications of our nuclear policy seriously and give the due importance to renewable energy sources." I concluded, "There is a great danger of our energy policy becoming the captive of the nuclear technological elite. Our national energy planning and our military and defence interests would be better served by developing solar technology."

In the Indian context, this was the first ever critical evaluation of nuclear power in relation to solar energy, and these articles became basis for active campaigning for renewable sources of energy. Until then, the Indian government had made no move to spend anything on research and development in solar energy.

ALARM BELLS START RINGING

By early 1981, Mrs. Gandhi was back in power and her anti-people and undemocratic style of governance was evident in a study released by the Press Council of India in July 1981. It stated that two of the wings of the government, namely the legislature and the judiciary, functioned in the open but "the executive does its business in its secret chambers to which the people have hardly any access." In the name of "national security" and "public interest" any information could be denied to the people. The condition relating to the nuclear energy department was even worse due to its sensitive nature.

In addition, I noticed secret and close linkages between big industrial establishments and the ruling political elite, particularly the prime minister's family members and the financial business interests in the country. In my investigation I was venturing into a sensitive area. But being a naive academician, I believed that by exposing the secret deals and unscientific nature of nuclear enterprises I would be able to reform the system. Instead, as I systematically advanced in my investigative exposures of nuclear industrial-cum-political operations, I invited the attention of the secret agencies of the state. Besides articles and interviews appearing in the newspapers, I was preparing the first book-length study of the nuclear programme in India in spite of all the official restrictions.

In December 1980, I visited England and met Right Honourable Tony Benn, who had been the Energy Secretary from 1974 to 1979 and who was known for his critical view on nuclear energy. During long discussions with Benn I learned about the secret functioning of nuclear establishments and received some important tips for comparative analysis of nuclear establishments in India and abroad.

To make a critical study of any government policy is not an easy task, especially when one investigates activities relating to strategic importance. The state acts as if it has something to hide from its own people, and I confirmed with Benn that most policies are conceived and executed without the knowledge of the citizens. More often than not, even legislators do not know the official secrets. In India the citizens have no legal rights to information and the situation is still worse as there exists no group such as a "Union of Concerned Scientists" or "Society for Social Responsibility in Science."

I soon realized my personal responsibility: I was equipped to undertake a critical study of nuclear power and was in possession of information about the working of the Indian nuclear program. In an atmosphere of the official secrecy, even if my investigations were inconclusive, I felt it my patriotic duty to offer a critical examination of the nuclear establishment for public scrutiny. I believed that so long as the nuclear policy decisions of the government were not subjected to independent scrutiny, proper understanding of its objectives would not be realized, and responsibility and accountability would not be properly attributed. In the total absence of information and critical evaluation, no recommendations for reforms could be offered.

In early 1981, while finalizing my book for publication, I sought a meeting with India's top nuclear scientist, the father of India's atomic bomb, Dr. Raja Ramanna. On an earlier occasion, he had received me inside his official headquarters at the Bhabha Atomic Research Center and accorded due courtesies, as I was then sent there by Prime Minister Desai. In reply to my request for another meeting, Ramanna replied:

> Dear Prof. Dhirendra Sharma,
>
> Please refer to your letter dated March 16, 1981. As I have seen several of your articles, especially the one that appeared in the Manchester Guardian some time ago, it is clear that we have very divergent views on the development of atomic energy in this country. I also feel that these your articles have damaged the country's reputation abroad. In view of this, I feel that there is no point in having a discussion on this matter.
>
> With regards,
> Yours sincerely,
> (R. Ramanna)
> Date: 23 March 1981.

Ramanna was also then the Scientific Adviser to the Minister of Defence and later became the Chairman of Atomic Energy Commission.

I was to learn later that Ramanna, the Edward Teller of India, was by nature intolerant of criticism and more than once had suggested to the Prime Minister to "stop Sharma," if necessary, by arresting me under the secrecy provisions of the Atomic Energy Act. The campaign or conspiracy to shut me up had begun under the instigation of Ramanna in 1981.

But my investigative research continued. I was getting my chapters typed in pieces and in three copies: one copy I kept in my office at the university, another one I hid in my residence, and the third one posted to my son in England for safe keeping. But I was still naive and did not visualize how far my adversaries could go.

Until then, I maintained my academic posture and my criticism was directed against nuclear power policy. I was asking critical questions. If the government claimed 10,000 MW (megawatt) of nuclear electricity would be produced by 1990, the nation must be told its cost and whether the country had the industrial and financial resources to support such a program. I estimated that India would require 32,000 metric tons of heavy water for initial inputs in its reactors and about 2,000 metric tons of heavy water annually to run

44 CANDU reactors each of 230 MW capacity, plus about 50,000 trained personnel to run these atomic power stations. All this would cost the enormous sum of about 25,000 million rupees at the 1983 rate. In early 1980s, I declared that the country did not have the industrial or financial resources to produce 10,000 MW power even by the year 2000 .

By 1993, India's installed nuclear power capacity was just about 1500 MW, and the country is nowhere near the nuclear establishment's proclaimed targets of the 1980s. In fact, the official estimates now have been lowered to about 5000 MW by the year 2000. But it was not this criticism which disturbed the authorities. It was, in fact, my exposures of corruption, mismanagement, and the issues relating to secret deals and financial arrangements that invited the wrath of the authorities.

The government argued that nuclear weapons were bad, but that nuclear power can be used beneficially. But I saw the reality: atoms for peace and atoms for war were Siamese twins which cannot be separated. If supply of electricity was the aim, I questioned the government, why not give just 10 percent of funds to research and development for renewable energy? And as I carried the campaign against the whole nuclear strategy, the nervousness of the establishment became more and more apparent.

India's official spokesperson for its Defence Policy (sic) and then director of Institute for Defence Studies and Analyses, Dr. K. Subrahmanyam, launched a campaign for India to obtain nuclear weapons to deter threats to the country's security.[3]

Once again I alone challenged this call for the bomb which was rooted in jingoism and lacked any science policy and/or defense policy perspectives. I claimed that simply shouting that the enemy is holding the atomic bomb against us is not a serious "policy" statement. In formulating a national defense policy, one must examine national and international implications and industrial and financial ramifications. Making the bomb is a technological mission and does not constitute a national defense policy goal which should be safety, security, and stability with social and industrial advancement of the country. I retorted that "to make a few bangs is easier than to run industrial and economic institutions efficiently. It is even more difficult to provide the millions with the daily basic needs. But the bomb hysteria would divert our attention from the fundamental issues of building up a just and equitable society."

But the editor of the *Times* refused to publish my article or even a rebuttal letter. I now recognized the urgency to combat this bomb-cry which was apparently raised by the pro-bomb lobby with the approval of Mrs. Gandhi. In order to boost her popularity she imitated Mrs. Thatcher and encouraged jingoism. I saw the necessity to make an organized counter campaign to stop the bomb hysteria.

In order to alert citizens against the nuclear bomb and nuclear power, I organized the Committee for a Sane Nuclear Policy (COS-NUP) in June 1981. This was the first antinuclear organization in India. Under the banner of COSNUP, I prepared a statement signed by twenty-four prominent citizens including the late Madam Vijayalakshmi Pandit, former Secretary-General of the United Nations and the sister of Jawaharlal Nehru, Ms. Nayantara Sehagal, a noted novelist and the first cousin of Mrs. Gandhi, and a few eminent jurists, renowned writers, and journalists. In the statement released by COSNUP on June 28, 1981, we expressed "deep concern over the re-emergence of nuclear bomb lobby in India," and urged the government not to take a rash decision in favor of the nuclear bomb because diplomatic channels are open to the country. We advised the government to perceive the problem of security from the wider South Asian perspective. After a lapse of three months, the editor of *The Times of India* relented and published my rebuttal to K. Subrahmanayan's pro-bomb article as a lengthy letter.[4]

As I emerged as India's most vocal antinuclear campaigner, COSNUP became a movement. I toured various cities and towns, wrote antinuclear articles, gave seminar lectures, and organized marches. But my problem was not so much to explain radiation hazards to educated citizens who mostly understood English. The vast majority of Indians live in villages and 60 to 70 percent of them are illiterate. My co-activists were university students, mostly urban, and all of us had acquired a critical approach to nuclear power technology by reading material in English. The problem was how to explain radiation safety in local idioms to the villagers.

There were also problems of transport services, and lack of communication where no telephone facilities existed. We had made a propaganda video against nukes but due to lack of electricity any modern gadgetry had little usefulness in rural India. Activists from cities had also to experience the privation of washing facilities where there was no running water nor any public toilets. In the 1960s, I had participated in anti-Vietnam War demonstrations at Lincoln Memorial in Washington, D.C. It was not the same to organize a demonstration in India's villages. But I found doing it just as rewarding an experience. For the first time, a highly qualified and west-

ernized science policy critic was witnessing the Indian village realities. It was an enriching experience which strengthened my resolve
that most high-tech systems, particularly nuclear power, were not
appropriate for India and other Third World nations, where 70 to 80
percent of the population live in rural areas and lack even elementary
modern amenities. Without proper roads and communication systems, how would you evacuate a few million citizens in the event of
a nuclear accident?

Moreover, to have an attentive village audience was another
problem. They were accustomed to listen to bargaining over wheat
and sugarcane, or discussion about irrigation of their fields or even
the electioneering of political parties. But no one had come to them
to discuss scientific arguments or to give technological information
about safety problems in a CANDU reactor. I found it difficult to
explain scientific terms in the local village dialects. Radiation is
invisible and odorless, so how could I explain that it remains hazardous for twenty-five thousand years? The villagers could not
believe that I was telling the truth. How come no one else, no political leader, no prime minister or political party, told us about this
danger of nuclear power? Why is it only Professor Sharma? "Because
I am a professor," I persisted.

But the problem was different with the educated scientific community: most of them knew what radiation was and what the problems are in nuclear power. They consider it prudent not to oppose
the government policy as the state commands enormous powers of
patronage and punishment. In India, almost 95 percent of scientific
institutions and research grants come from the government. Hence
there existed no scientific autonomy, particularly in higher research
and educational institutions. Appointments to all top posts, including university heads and heads of research and development organizations, were made on approval of the Prime Minister. When I
approached the scientists at the Tata Institute of Fundamental
Research, the foremost institution in India, I was rebuffed: "no one
here is qualified to comment on the question of safety in nuclear
reactors." Internationally renowned astrophysicist, Dr. Jayant
Narlikar, told me that he was seeking a grant of ten million rupees to
establish his Institute of Astrophysics and so could not be bothered
with nuclear policy controversies especially since it was likely to
offend the Prime Minister.

My antinuclear campaign was picking up momentum but so
also were efforts to "stop Sharma." The main character in my confrontation with nuclear power in India was the lion of the Indian

Atomic Energy establishment, Dr. Raja Ramanna, the father of India's Pokharan explosion and the chairman of the Atomic Energy Commission, 1982–1987. For years he headed secret research as Director of the Bhabha Atomic Research Center (Bombay). Although he had asked more than once for my arrest under the secrecy provisions of the Atomic Energy Act 1962, a more efficient and benign procedure was adopted.

At that time, in 1981, Raja Ramanna was the Scientific Adviser to the Minister of Defence, Government of India. A review committee was appointed to "formulate a working programme" for the Centre for Studies in Science Policy, JNU. Ramanna himself headed the committee. I could foretell the outcome. After a few casual meetings with a few faculty members of JNU, on 7 February 1981, Ramanna gave his verdict that the Science Policy Centre should be closed down. He wrote:

A Nine Men committee was constituted by the Executive Council of the Jawaharlal Nehru University to advise the Vice-Chancellor on the need for a Centre for Studies on Science Policy and its activities. After considerable discussions, the Committee made the following recommendations:

1. That as a field of research there definitely exists subject which can be termed "Science Policy." In this field, studies could be undertaken on a number of topics, e.g., law of Seas, Science Education, Energy Option etc. It may also include foundational areas like Philosophy of Science, Sociology of Science, History of Science and Technology and Psychology of Science. For this purpose, it does not seem quite necessary that Centre for Studies in Science Policy should exist, but the research worker should be able to move freely, in various related departments where they can discuss the issues with experts in the concerned overlapping fields of knowledge.

Ramanna disregarded my request for a meeting with the committee and refused to look at my course material and research publications. But his official report stated that I was abroad on the day called for discussion, and recommended that I should be transferred to "any other center willing to accept him" or sent out of the university, if necessary!

I strongly opposed the Ramanna Report in the Academic Council and in the Boards of Advanced Studies and warned my colleagues that if they accepted the report, it would set a precedent, and the government in future could close down any other center whose faculty might express critical opinion on government policies. JNU Academic Council formally thanked Raja Ramanna for painstaking efforts but the report was shelved. Yet, for all practical purposes the Science Policy Centre was placed in deep freeze. The Centre was not permitted to admit new students or supervise any doctoral candidate.

Around this time, the late Dr. Y. Nayudamma became JNU Vice-Chancellor. He was himself concerned about science policy issues and was personally known to me. In fact, he had written the foreword to my volume *Science and Social Imperatives* (1976). Those were the days of the Emergency, and he was then Director-General of the Council of Scientific and Industrial Research. Now, being my vice-chancellor, he invited me to restructure the Centre and in view of my senior colleague retiring within a few months, suggested that I should plan for taking over responsibilities of the Science Policy Centre again. Nayudamma was a scientist of integrity and held independent views. He was one who often disagreed with the Prime Minister. Mrs. Gandhi respected him for his courage and appointed him Vice-Chancellor for a term of five years. But as he worked to clean up the JNU administration, in less than two years he was forced to resign. The man who succeeded Nayudamma, Dr. P. N. Srivastava, though a scientist of some repute, was a climber at best.

Srivastava, while a professor of biology at the university, had written a research paper supportive of the official line that low-level ionizing radiation is not hazardous (and indeed in low doses is good for health) and that nuclear energy is safe. On the question of the appropriateness of nuclear power, I confronted him in a national debate before a group of scientists and antinuclear activists in the "science city" of India, Bangalore.

As the new Vice-Chancellor, Srivastava ran a totally secretive and repressive regime in the university. He was apparently in league with Mrs. Gandhi's power brokers as was evident from his later posting to a ministerial post in the Planning Commission. The Students Union in JNU was avowedly anti-Prime Minister and pro-Opposition Party. Srivastava banned student political activities and ruthlessly crushed all student agitation on campus. Hundreds of students were mercilessly beaten and arrested and a host of them were expelled from the campus, leaving their academic careers in ruins. The message was clear that the new Vice-Chancellor would act as the henchman of the Prime Minister.

THE FINAL BLOW TO MY ACADEMIC CAREER

The 1982–1983 period was most productive for me from many aspects: the antinuclear campaign was at its height and my articles were appearing in national newspapers. My most controversial book, *India's Nuclear Estate*, was released in May 1983.[5]

Dr. Raja Ramanna's appointment as Chairman of Atomic Energy Commission was announced on 6 August 1983. In a scathing criticism of Ramanna's policy, I wrote: "since he has been affiliated with, and is known for his keen interest in, advanced nuclear and defence research, the new chairman is likely to push the country towards an open nuclear weapons policy. If he does this he will receive support from populist politicians and the powerful military-industrial complex in the country."[6] I concluded by demanding more open and democratic decision-making processes in the Department of Atomic Energy. I also pointed out the significance of the announcement of Dr Ramanna's appointment occurring on Hiroshima Day.

In November, I criticized the Atomic Energy Department for constructing an atomic power station in a high seismic zone, one hundred miles from New Delhi at Narora, situated only fifty-six miles from the active Moradabad fault of the 1956 earthquake. Based on my study of official secret reports, I claimed that the Narora site was never cleared by the Site Selection Committee; it was a political decision of Mrs. Gandhi who offered the project to upset the popular base of her powerful political opponent Chaudhuri Charan Singh.[7]

At the end of November that year, Mrs. Gandhi hosted the Commonwealth Heads of Government Meeting (CHOGM) in New Delhi. I took the occasion for launching another campaign: "let all the [British] Commonwealth countries collectively declare themselves nuclear-free territories." Under the banner of COSNUP, I organized a petition signed by some two hundred eminent citizens including members of parliament, professors, lawyers, architects, editors, and journalists, which appealed to the Commonwealth leaders to provide "moral courage and leadership" to the world by taking the first step towards collective nuclear disarmament. My petition urged CHOGM to assert that no Commonwealth state would enter into agreement with any other government for stationing nuclear weapons or for their possession or production. The COSNUP appeal decried global military expenditure which had then reached the alarming level of $800 billion a year. The petition warned that radiation does not discriminate between friends and foes, and urged,

"There is no necessity to add a single nuclear weapon to the stockpile and safety and security of the world cannot be brought closer by nuclear weapons. Therefore, we call upon CHOGM to affirm a Commonwealth Nuclear Weapons Policy, which collectively renounces the testing, production and use of nuclear weapons by the Commonwealth countries, bans installation and stationing of nuclear weapons from all Commonwealth territories and assures never to use or support the threat of deployment of nuclear weapons to resolve international conflicts."

Indian newspapers welcomed such an appeal and *Indian Express*, the most popular national daily, editorially supported COSNUP's appeal to CHOGM.[8] I took the appeal to the Secretariat of CHOGM so that it could be included in the agenda. I sent my request to a few embassies but I was told that only a head of the government can insert an item in the agenda. Except for the Australian Minister for Disarmament, no one was willing to discuss my innovative approach to nuclear disarmament.

The two most powerful individuals in the Government of India, Mrs. Gandhi and Dr. Ramanna, were reported to be upset on my latest antinuclear salvo which I thought would embarrass the Conservative Government of Mrs. Thatcher more than Mrs. Gandhi. The Vice-Chancellor of JNU, Srivastava, saw the opportunity to please the Prime Minister. Immediately after the CHOGM meeting, when the Executive Council of the University met on 6 December 1983, he got a resolution passed "to agree to transfer Dr. Sharma from the Centre for the Study of Science Policy, School of Social Sciences to the School of Languages with immediate effect." Who could oppose such a brilliant move to make the Prime Minister happy?

In my reply to the Vice-Chancellor I pointed out that only a few days earlier, when I had called on him to discuss some work of the Science Policy Centre, he had given me no indication about any possibility of my transfer. And only a few weeks before that, I was invited to plan the future development of the Centre and was assured that I would be promoted to full professor and take up the chair of the Centre. I asserted that if the decision was due to "any academic compulsion, the matter could have been discussed with me, as I alone to be affected by the decision. I am the seniormost faculty in the Centre and the order of transfer at this stage is evidently to stop my chances of promotion" in the field of science policy.

I pointed out that there is no provision in the Rules of the University which allows an arbitrary transfer of a faculty member, after having confirmed him/her in a center for more than ten years. One sympathetic official of the university passed on to me photocopy of the Rules relating to transfer within the university which read "The transfer of faculty members from one Centre to another may be made with the written concurrence of both the Centres as well as of the faculty member concerned."[9] But all my appeals and petitions to the Vice-Chancellor, Registrar, the Minister of Education and the JNU Teachers Association remained unacknowledged.[10]

EVIDENCE OF AN ANTI-SHARMA CONSPIRACY

In December 1983, I was transferred and the Centre was closed. But the government in a reply to the Parliament promised that the Centre would be reopened within a year. A decade later the Centre still remains closed. For more than a decade now, science policy, as an academic discipline in India, has been dead. During my tenure, I had developed M.Phil. and Ph.D. programs and there were eight research fellowships for Ph.D. candidates. Perhaps I was incompetent to run a Science Policy Centre. But what I learned about the official approach to my confrontationist attacks on powerful individuals was alarming.

After my transfer from the Centre, rumor was spread that I was a foreign agent, specifically "a CIA agent." Because of this character assassination and whispering campaign, within and without the university, for all practical purposes I became persona non grata. In the university no faculty would speak to me and activist students stayed clear of me. There was nothing I could do effectively in isolation. During these days, I once walked into a government scientific department to visit an old friend of my family. He did not speak to me, and after some time when I questioned his behavior, he blurted out that "we have been told that you were a CIA agent." I left hurriedly in disgust.

I personally knew a Nehru, a first cousin of Mrs. Gandhi, Mr. B. K. Nehru, Director of Nehru Memorial Funds and executive head of many other foundations relating to the dynasty. (He is different from his brother of the same initials. The other B. K. Nehru, more intelligent and forthright, was India's High Commissioner in the United Kingdom and was removed from Governorship of Jammu and Kashmir state for his critical stance vis-a-vis Mrs. Gandhi.)

Sometime in summer 1984, by chance I met the Nehru in a restaurant, and asked his help to present my case to the Minister of

Education, who was his and Mrs. Gandhi's aunt. He said "there should be no problem" to arrange it. After a few days he asked me to see him in his office at Teen Murti House. His face was burning red and, without the usual pleasantries, he showered me with condemnation. I cannot quote him here verbatim because I did not tape him nor could I take notes. If only I had had a premonition of what I was going to receive. The Nehru said something like this:

> We Nehrus have ruled this land for one hundred years, since the first Nehru [Motilal, the father of Jawaharlal Nehru] became president of Indian National Congress in 1920s. We are the masters here whether you like it or not. Go to any city or town and you will see a park, road, school, or hospital named after Nehru family. Turn any stone and you'll see Nehru engraved in every mountain. . . . Of course, you are a great scholar and you have right to hold your views. That is your democratic right. But who cares for your constitutional rights in this country? It is we Nehrus who grant you that right. But if we say No, No, then that goes in this country. If you don't like it, you are free to go. Your family lives abroad, in Britain and in America. Why don't you too go away? I advise you to leave India, as you will do better there. In this country we shall not allow you to teach any science policy. If necessary we would close down the whole university . . . if necessary.

The Nehru was fuming. After a few minutes' pause he became a bit composed, and slowly tried to explain the background of his frustration.

> You see, it is not a simple academic freedom issue here, as it is in America or in the United Kingdom. They [in the Ministry of Education and in the Department of Security] have a huge file on you and of your writings and reports of your speeches. You have one refrain that Mrs. Gandhi is anti-people and that the nuclear program is designed for evil purposes. At best one twists the tail of a lion. But you have placed your head in his mouth and the lion had crushed your head, smashed you. . . . You have been challenging the power and the power has responded. You can do nothing to us. . . .

As I left the Teen Murti House—former official residence of the late Prime Minister Jawaharlal Nehru—I noted that whole acres of the palatial estate of the Indian government have been taken over *free* for a personal Nehru Memorial Foundation. While thanking him for such a frank talk about the system of our democratic India, while leaving Teen Murti I heard myself saying "I'll not be ruled by the dynasty and leave India. I shall confront you with all my scientific knowledge and political wisdom."

Within the university circles I felt dejected, not because my Center was closed and I was transferred but because I was labeled a CIA Agent. In the 1960s, I taught at Michigan State University in East Lansing, and then I actively opposed Vietnam War and joined in civil rights marches. In the United States I was accused of being a red—a Maoist. Consequently, the Fulbright-Heys Research Fellowship awarded to me in 1969–1970 was withdrawn by the U.S. Department of Health and Education under the intervention by newly elected President Richard Nixon.[11] Now, in India, in my home country, I was supposed to be a CIA agent!

I decided to mobilize international support, particularly since in India no scientist or intellectual came forward to defend my academic freedom. I had known Noam Chomsky from my U.S. activist days. I had also been in touch with Tony Benn in the United Kingdom. I also wrote to Professor Paul Sweezy, who had suffered under the McCarthyist repression in the United States for his liberal economic theories. All of them and many others sent their protests to the university and Chomsky in a lengthy letter to the editor of *The Times of India* wrote:

I have known Dr Sharma for almost 20 years. He was a courageous and effective participant in the American anti-war movement, and has since done important and highly-valued academic work in the area of science policy while continuing with his engagement in defence of civil and human rights in India and elsewhere in the world. His active opposition to the Indo-China war apparently cost him a U.S. government research fellowship in the year 1969–70. No stranger to controversy, Dr Sharma has always conducted himself with great honour and integrity, both in his scholarly work and his activities in connection with problems of freedom and justice.

It is hardly necessary to stress that the very existence of a free university depends on vigilant defence of the right of scholars to draw the conclusions to which their research leads them without fear of punishment and discriminatory action by higher authorities. I trust that this decision [of his transfer] will be revoked and that Dr Sharma will be afforded the opportunity to continue his important work unhampered.

Chomsky's letter was dated 26 March 1984. Mr. Girilal Jain was then the editor of *The Times*. He was known for his pro-Mrs. Gandhi policy. In a letter to me dated 2 May 1984, Mr. Girilal Jain curtly stated that he was "unable to publish Prof. Chomsky's letter."

Meanwhile, Tony Benn and concerned scholars, including editors Les Levidow and Robert Young of the London-based journal *Science as Culture*, sent protest letters to the editor. Eventually, on 18 May 1984, the editor reluctantly published the protests in the letters column of *The Times of India*.

How could these radical thinkers of the west be defending a CIA agent? Noam Chomsky's and Tony Benn's letters had a sobering effect upon self-styled radical intellectuals of India. The government intelligence services must have goofed up somewhere. But interestingly no JNU faculty member, no scientist, no political party, or prominent leader in India came forward to defend my academic rights. The JNU Teachers Association lodged no protest at the violation of university rules of the transfer of a faculty member from one centre to another which required the written consent of the faculty member concerned.

There was, however, one exception: the former Foreign Minister in the Janata Government (1977–1979), a senior parliamentarian, and the leader of the Opposition in Lok Sabha (Lower House), Honourable Atal Behari Vajpayee, addressed a protest communication to the Chancellor, JNU, Dr. D. S. Kothari—another known yes-man to Mrs. Gandhi. Mr. Vajpayee on 8 October 1984, referring to the rule of transfer, asked the pertinent question:

> What were the reasons that led to the transfer of a teacher from the Centre, in which he was appointed and confirmed, about a decade after his appointment?
> It appears from the circumstances of the case as reported in the Press that Dr. Sharma has been transferred because of his views—his critical examination of India's nuclear programme. If so, I am sure you will agree that this is a serious matter.

He referred to Noam Chomsky's letter about academic freedom and asserted that "dissent and debate, on public policies in particular, is an essential element of the democratic way of life. As a member of the Lok Sabha from Delhi, and as a member of the Court of the JNU, I feel particularly concerned about the case." He urged reconsideration of my transfer.

But for all practical purposes JNU's Science Policy Centre was closed and there was no institution of higher learning and research in the country which could offer me teaching and research facilities in science policy. But my antinuclear campaign continued with better media coverage in the country and I enjoyed greater international recognition.

POSTSCRIPT

On 6 June 1992, when I retired from the university unceremoniously, it took me six months to get all my dues from the university and I do not remember how many times and how many administrators I had to visit personally in order to complete unnecessary formalities. But this was not harassment. Within six months of my retirement, the Science Policy Centre at JNU was reopened with new faculty appointments.

But in my efforts to build up a critical perspective towards scientific and technological policies in a country where it was blasphemy to criticize those in power and where it is not customary to be critical of the government science policy, I have some successes to record. Following the publication of my book *India's Nuclear Estate* in 1983, in which I criticized and made constructive suggestions for reforms, the following initiatives were taken by the Government of India:

1. In 1984, a small unit under the name "Atomic Energy Regulatory Board," with a few rooms and furniture inside the Department of Atomic Energy, was created. It is still only a Board, under the Atomic Energy Commission. And though it is not an independent commission, over the last few years it has better office facilities and about fifty personnel.
2. In 1984–1985, the Government of India established an independent Department of Non-Conventional Sources of Energy to encourage research and development in renewable resources of energy. Until then, nuclear power was considered to be the sole contender for energy future and zero funds were made available for renewable energy sources.
3. In 1984, following the criticism I made in my book that India's Comptroller and Auditor-General did not look into the accounts

of Atomic Energy Department, a special cell was formed by the Auditor-General of India to investigate and do some accounting of atomic energy, space and defense research, and development departments. This has now become a regular feature and, even though not completely satisfying, a beginning has been made to look into the financial affairs of the atomic energy and other secret science and technology departments of strategic importance which used to be free of mandatory accounting of government departments by the Comptroller and Auditor-General of India.

At the end of my story of confrontation with the state power, I have the satisfaction that I did not bend or break throughout the period of my struggle. During all those critical years three things sustained me.

First was repeated confirmation that I was fighting for the right cause. Access to scientific literature, and my constant exchanges with science policy critics abroad, proved great help. My campaigning against the dynastic government and for the antinuclear movement, as I perceived them, was part of global scientific-democratic movements of the twentieth century in which radical forces of scientific values offered revalidation of socio-political systems.

Second, temperamentally, I perform best when in confrontation with powerful authorities. And in this I was very much inspired by Bertrand Russell who laughed at the intellectual weaknesses of rulers and heads of government departments. Basically I was opposed to authoritarianism and believed that in the final analysis "Truth must win."

Third, I was sustained by my life-partner Nirmala, my wife, who, at every critical juncture, stood by me. She nursed my determination not to give in or compromise with unjust pressures or to succumb to the temptations of grants, position, or promotions. During the Emergency, she provided shelter to my underground political activists who were hounded by the secret police of Mrs. Gandhi. She was always there when I needed assurance that the path of confrontation I had chosen was for a just cause.

If I had my time again, I would confront the challenges with even greater vigor.

NOTES

1. See Dhirendra Sharma, ed., *The Janata (People's) Struggle: The Finest Hour of the Indian People*, New Delhi: PSA Publication [M–120

Greater Kailash-I, New Delhi 110048, India] (1977), which includes under-ground documents, resistance literature, and correspondence relating to advent of the Janata (People's) Party, which challenged Mrs. Gandhi's dynastic government.

2. Dhirendra Sharma, "Time to Move away from Nuclear Power," *The Times of India* (21 and 22 August 1980).

3. K. Subrahmanyam, "A-Bomb the Only Answer," *The Times of India* (26 April 1981).

4. Dhirendra Sharma, "No Bomb, Please," *The Times of India* (5 July 1981).

5. Dhirendra Sharma, *India's Nuclear Estate*, New Delhi: Lancer International (1983). See also Dhirendra Sharma, ed., *The Indian Atom: Power and Proliferation: A Documentary History of Nuclear Policies, Development and the Critics: 1958–86*, New Delhi: Central News Agency (distributor) (1986); Dhirendra Sharma, "India's Lopsided Science," *Bulletin of the Atomic Scientists* (May 1991): 32–36.

6. Dhirendra Sharma, "Dawn of a New Atomic Era?" *The Hindustan Times* (1 September 1983).

7. Dhirendra Sharma, "Narora: Threat To Ganga," *The Hindustan Times* (12 November 1983).

8. *Indian Express* (23 November 1983).

9. Resolution no. 4. 1/EC/17. 9 (1979).

10. For an account of my transfer see Brian Martin, "Nuclear suppression," *Science and Public Policy*, vol. 13 (December 1986): 312–320.

11. For details see *The State Journal*, East Lansing, Michigan (3, 5, 6 and 10 June 1969) and the autobiography of the then president of Michigan State University, Walter Adams, *The Test*, New York: Macmillan (1971), pp. 152–158, "The outside agitators."

8

Conclusion: Learning from Struggle

Each of the six preceding chapters has described a challenge to a powerful establishment. Since I invited most of the contributors independently of each other, they do not necessarily agree with the positions or methods adopted by the others. What they have in common is the experience of challenging the experts.

Drawing on these case studies, this final chapter has three main points. The first is that it is incredibly difficult to dent an establishment position. A second important message, in direct contrast, is that even a few critics can make an enormous difference. The third message is that most people are excessively acquiescent, and that more should be done to increase the possibilities of debate.

THE POWER OF ESTABLISHMENTS

Establishment experts are in a powerful position. Typically, they have superior numbers, prestigious positions, high credibility with the media and the public, control over professional and academic journals, and links with powerful groups. Faced by a challenge, their usual initial response is simply to ignore it. Harold Hillman, for example, published many papers critical of biological orthodoxy, but for many years it appeared that no one took any notice. Only an establishment can get away with this. The standard view is so completely taken for granted that critics are assumed to be misguided.

When a critique is "ignored," often there is suppression involved, such as the prevention of publication in key journals or a refusal to review writings by critics in suitably prominent fashion. In other words, to say that the critique can be "ignored" often means that suppression is working in a quiet, behind-the-scenes fashion. If, in spite of this, critics become too noisy, too credible or too influential, then they are liable to be suppressed in a more overt and heavy-handed fashion, for example by personal attacks on the dissident.

This is a pattern found over and over again in challenges to expert establishments. For example, when Hillman simply published his critiques of standard methods in biology in scientific journals—and often that was hard enough to achieve—other scientists could simply decline to take notice. But when he issued challenges in prestigious scientific meetings or obtained publicity in the media, then "quiet" suppression was not enough. He was met by deceitful "refutations" of his views, bureaucratic slights and "mistakes" that hindered presentation of his views at scientific meetings, and a major threat to his laboratory and his academic career. His experiences are replicated repeatedly in other challenges to establishments, though with innumerable variations depending on the situation and issue.

If being ignored or being suppressed were the major problems in confronting establishment experts, this would not be such a difficult business. There is something more involved: vested interests behind the establishment position. Indeed, vested interests are crucial in making a position into one called an "establishment."[1]

For example, Edward Herman confronted not just a few establishment experts on terrorism but also an entire political system that benefits from the orthodox position on terrorism. This includes the U.S. government agencies and businesses—including spy agencies, diplomatic corps, and multinational corporations—that want to keep on good relations with murderous regimes, and so prefer that the label "terrorist" be reserved for something else. This establishment provides the sponsorship for intellectuals who defend the orthodox view. All in all, there is enormous material benefit for supporting the standard view on terrorism versus little reward, and possibly a lot of lose, by questioning it.

The link between experts and vested interests is even more obvious in the case of the nuclear establishment in India. Dhirendra Sharma, by challenging nuclear policy openly, came up not just against nuclear experts and bureaucrats but also against a close-knit political and economic elite with a stake in nuclear developments. Indeed, the nuclear scientists and engineers would not have been a formidable force without their connections with some of India's most powerful figures.

Because the power of establishments is so great, many of the most effective critics come from the outside, where they are less tied to the main professional bodies or patronage system. For example, Sharon Beder was not a Water Board engineer, Mark Diesendorf was not a dentist or doctor, Edward Herman was not sponsored by the U.S. government, Michael Mallory and Gordon Moran were not from Siena or even Italy, and Dhirendra Sharma was not a nuclear scientist. Ironically, this independence of vested interests is often criticized as a lack of proper credentials or expertise. When an expert establishment holds sway, being coopted by the official patronage system actually adds to an expert's credibility.

But it would be too crass to attribute the strength of the establishment simply to money, jobs, and power. These are the material foundation for a position, but to be really effective, psychological commitment must be involved. In every case, the establishment has a comprehensive world view to which leading figures are intensely committed. Most of the establishment experts believe that the critics are wrong, misguided, and even dangerous—in fact, in the view of many, sufficiently misguided and dangerous to warrant the various actions taken against them.

In the case of terrorism, the establishment experts believe they are addressing the greatest threats to peace and freedom. In the case of sewage, the establishment engineers believe that their approach is the only effective way to proceed. And so on through every case study.

An establishment based on cynicism would not last long. Most people seek to mesh their beliefs with their actions. An establishment position heavily based on conscious deception or consciously unfair behavior would quickly lead to defections. Personally, even though I may consider the behavior of some experts to be underhand or reprehensible, nevertheless I have always considered them to be sincere—though that sincerity may be based in a worldview quite contrary to mine.

No doubt some establishment experts consciously lie in order to defend orthodoxy, but this should be put in context. The power of rationalization is enormous, and so it can be expected that most experts (like other people) are likely to adapt their beliefs to a world view that serves their self-interest. Furthermore, for some, lying occasionally may be a means to a greater end, namely defending a position they *know* is best.

It is the combination of vested interests and commitment to a worldview that makes the establishment position so hard to dent. The material factors (the vested interests) provide the basis for power and the mental factors (the worldview) provide the willingness to use the power. Critics often begin by thinking that if they can find and demonstrate holes in the arguments used to defend orthodoxy, then its position will collapse. But picking holes in arguments neither changes the vested interests nor, in most cases, undercuts the prevailing worldview. Furthermore, if the critics only occasionally get a chance to be heard, the establishment position may be accepted purely through repetition: it is so often stated that it seems to be "common sense." Is it any wonder that critics can so easily be ignored?

THE POWER OF CRITIQUE

The second message from the cases in this book is that a small number of critics—sometimes just one—can make an enormous impact. Indeed, suppression of dissent is a signal that dissent can make a difference. If there is no dissent, suppression is not required.

A crucial part of the critic's effectiveness is strong arguments. In every case, the critics have begun by mustering powerful intellectual attacks on the orthodox position. This is not just a matter of moral conviction, of standing up and shouting "You're wrong!" in the face of the establishment. No, the secret of every successful critic is good arguments, based on collecting information, carrying out careful analyses, preparing well thought-out written and verbal presentations.

All of this requires a lot of hard work. Reading the case studies is not likely to give a full sense of the amount of work involved. A chapter recounting days in the library would hardly be interesting, and every author inevitably emphasizes the more dramatic events in the story. But without the long hours of study and preparation, the highlights would never have occurred.

There seems to be a contradiction in my argument: I said that establishments are held together psychologically through a worldview and yet I'm saying that the arguments of critics can be effective. How can arguments puncture the worldview? The resolution to this apparent dilemma is that the arguments of critics are most effective in convincing third parties, namely people who are not part of the establishment position. This might be politicians, media, experts in related fields, or members of the "general public."

Both establishment experts and critics are engaged in a contest over loyalties. The establishment, by definition, has the advantage of the loyalties of the most powerful and authoritative experts. The establishment, to maintain its power and authority, has to keep it this way. The critics can make inroads by winning over a few recruits, for example from new or marginal members in the orthodox camp, from groups that are not officially part of the establishment, or occasionally even a convert from the mainstream of orthodoxy. In all these cases, arguments can be effective, though they are not enough on their own to win the day.

The visibility of just a few critics turns unanimity, or at least the appearance of complete agreement, into a debate. From the point of view of outsiders, this is enormously important. Instead of the orthodox view being taken for granted, it becomes simply one point of view. This weakens the position of the establishment dramatically. For example, Sharon Beder describes the crisis at the Sydney Water Board when sewerage issues became of widespread interest, with the media reporting critics as well as establishment views.

None of this would make any difference if the critics had only arguments. To be effective, these arguments need to be linked to interest groups, in the same manner that establishment experts are linked to vested interests. For example, the arguments of critics of Sydney sewerage policies were taken up by environmentalists and beach-goers. For establishments, critics alone are not much to worry about. It is their potential to aid and help to mobilize interest groups that is a real threat.

It is for this reason that critics are likely to be attacked. If the credibility of the critics can be undermined, then their threat to establishment legitimacy can be minimized. Each case study has plenty of examples of attempts to discredit dissidents, such as accusing Dhirendra Sharma of being a CIA agent, and to suppress them or their work, such as forcing Sharma to move from his science policy post. But suppression, however damaging it may be for the person or position attacked, can also be counterproductive for the attacker. Suppression can backfire because it is perceived to be unfair. Many people believe, in principle anyway, in the value of open debate. When they are informed that debate is being suppressed, they may become more sympathetic to the suppressed position. Michael Mallory and Gordon Moran give some excellent examples of this phenomenon.

The key players in these confrontations include the mass media. If the establishment is unquestioned, there is no story. Even a single critic who has sufficient credibility, such as the appropriate credentials, turns the situation into a debate that is, therefore, newsworthy. The media has played a big role in the disputes over sewerage, fluoridation, the Guido Riccio, and nuclear power, among others.

The involvement of the media is especially potent in cases where establishment experts normally operate in the background without scrutiny, as in the case of most scientists and engineers. These experts generally detest media coverage. Ironically, it can force them to become more media savvy, as in the case of the Sydney Water Board, which has launched publicity campaigns defending its policies. But at least this visibility also makes the issue more available for debate than before.

The case of terrorism shows a different pattern. Here, a central feature of the establishment position is use of the media to inculcate the orthodox view of terrorism and to authenticate the establishment experts. In this situation, it is not a simple matter for the media to "open up the debate" because a key part of the "debate" should be the ways in which the media shape perceptions of terrorism and of expertise about it. In this situation, the "alternative" media, including community radio and small independent magazines and newspapers, become more important.

As well as the media, there are some other key players. Social movements are vital in a number of cases: the environmental movement in the case of sewerage, the antifluoridation movement, the peace and Central American solidarity movements in the case of terrorism, and the antinuclear movement. Movements are eager recipients and disseminators of work done by critics. Also important are quiet sympathizers or facilitators of debate. This can include an editor who decides to publish an article by a critic in a journal normally monopolized by the orthodox, or the organizer of a conference who makes a special point of inviting critics as well as defenders of the establishment. Some of these individuals may sympathize with the critics but be unable to make a public stand; others simply believe in the value of open debate. In either case, their efforts, while seldom dramatic, are vitally important in opening up the issue.

To become an effective critic of establishment experts, I think the following are crucial:

- lots of hard work, in order to understand the issues and develop the critique;

- a commitment to accuracy, since critics are more easily attacked and discredited by errors than are establishment experts;
- a willingness and ability to take the arguments to broad audiences, especially through the media;
- persistence;
- courage to disagree with peers and to continue in the face of attacks; and
- a secure livelihood.

The last item, a secure livelihood, is far from trivial. Many potential critics are deterred because of worries about their jobs. The most secure position is one completely independent of the establishment being confronted. Edward Herman is closest to this situation. The most risky position is to attack the establishment that provides one's livelihood. Harold Hillman is in this category and found that even academic tenure was insufficient protection.

In summary, even a single critic can do a lot against a seemingly impregnable establishment. By developing cogent arguments and raising them wherever possible, an undisputed orthodoxy can be turned into a debatable issue. In this, the involvement of a range of individuals and groups is important, including social movements, the media, and inside sympathizers. The critic is likely to encounter various forms of suppression but, on the other hand, may be supported by neutral parties who believe in fair play. Finally, in order to become an effective critic, there is a need for hard work, accuracy, taking the arguments to wide audiences, persistence, courage, and a livelihood. There is certainly room for more to join their ranks.

AN ACQUIESCENT SOCIETY

In any study of critics versus establishments, there is a great temptation to focus on the personalities of the critics. This might be to discredit them by pointing to psychological quirks or to praise them as exceptional human specimens. Of course, personalities are fascinating and worthy of study, but I think it is just as important to ask why there are so few critics. In Western liberal democracies there is much rhetoric about the importance of individual freedom and autonomy, but the reality is that most people are highly reluctant to openly challenge their superiors or even their peers, whether in corporations, governments, professions, or whatever. Most people are quite comfortable conforming to the prevailing views.

That in itself is not necessarily a bad thing. What is worrying is the limited support for open, vigorous debate. The contributors to this book each think that their position is correct, but they would not want a dictator to enforce their views by fiat. Rather, their wish is that the issues be debated openly and fully, allowing individuals to make up their own minds. For any society that calls itself free, this seems like an obvious and essential requirement. Honest debate often generates new positions and insights which are not available to any individual or group working within its own framework. Debate thus is essential to any society that aspires to be creative in the widest sense.

As I've indicated, becoming a critic requires a considerable commitment, and is certainly not for everyone. Furthermore, many people are quite satisfied with either the establishment position or a particular alternative position. But there is still an important role for those who do not want to join the debate as participants, and that is to be supporters of debate itself. Journalists can do this by seeking out minority viewpoints. Editors of newspapers and journals can do it by being more receptive to submissions by critics, or by setting up special for-and-against columns. Teachers can promote debate by collecting materials by critics to counterbalance establishment experts, and by inviting speakers from both sides of issues. Indeed, anyone can promote debate by organizing a public meeting with speakers from different viewpoints or having a meeting of friends to discuss conflicting writings.

Promoting debate sounds easy in principle but it can be difficult in practice. In most bureaucracies, including corporate, government, church and trade union bureaucracies, suppression of dissent is the usual pattern.[2] Even within social movements such as the feminist or environmental movements, which themselves are engaged in challenging establishments, internal criticism is often unwelcome.

To support debate is often seen a tantamount to supporting the critics, since debate gives the critics a platform that the establishment would prefer to deny. But this is no excuse for acquiescence. Without debate, no position is worthy of the unreserved support that establishments come to expect. That is precisely why it is necessary for more people to learn how to confront the experts.

NOTES

1. In some cases, the main body of experts is opposed to the primary vested interests, as the case of nuclear winter scientists versus the military

establishment (see Brian Martin, "Nuclear winter: science and politics," *Science and Public Policy*, vol. 15, no. 5 (October 1988): 321–334). This creates a somewhat different dynamic to the one presented in this book.

2. Deena Weinstein, *Bureaucratic Opposition*, New York: Pergamon Press (1979).

Contributors

Sharon Beder worked as an engineer before undertaking a Ph.D. at the University of New South Wales on the history and politics of sewage disposal in Sydney. Her work was instrumental in undermining the credibility of the Water Board and its engineers and bringing the debate about sewage policy onto the front pages of newspapers. Her book *Toxic Fish and Sewer Surfing* appeared in 1989 and was an immediate success. In 1992, she joined the Department of Science and Technology Studies at the University of Wollongong. Her social research has included special attention to the role of the expert. Her latest book is *The Nature of Sustainable Development* (1993).

Mark Diesendorf in 1994 joined the Australian National University to teach in the interdisciplinary Human Ecology Program. In the early 1980s, he was a principal research scientist in the Commonwealth Scientific and Industrial Research Organization, where he became leader of the Applied Mathematics Unit. More recently, he was senior research fellow and head of the Medical Services Unit, Australian Institute of Health, and then Coordinator of the Australian Conservation Foundation's Global Climate Change Program. He has been an active member of several community and professional groups, including secretary of the Society for Social Responsibility in Science, president of the Australasian Wind Energy Association and committee member of the Australian and New Zealand Solar Energy Society. He has published over seventy scholarly papers, edited four books, written many popular articles, and broadcast a number of radio talks and interviews.

Edward S. Herman is Professor Emeritus of Finance, Wharton School, University of Pennsylvania, where he specialized in corporate and finanacial power and regulation. For some years he also taught a course on Media Bias in the Annenberg School of Communications at Penn. He is the author of numerous books on economics, politics and the media, including *The Political Economy of Human Rights* (with Noam Chomsky) (Boston: South End Press, 1979, 2 volumes), *Manufacturing Consent* (also with Chomsky) (New York: Pantheon, 1988), and *Corporate Control, Corporate Power* (New York: Cambridge University Press, 1981), and two volumes devoted to the subject of terrorism.

Harold Hillman has University of London degrees in medicine, physiology, and biochemistry. He is Reader in Physiology and the Director of the Unity Laboratory of Applied Neurobiology at the University of Surrey, Guildford, England. He is also medical advisor to the Institute of Biological Psychiatry, Bangor, Wales. He has written five books and about 150 full-length publications on cell biology, neurobiology, and resuscitation.

Michael Mallory was educated at Yale University and received his Ph.D. in art history from Columbia University. He has taught at Brooklyn College of the City University of New York (CUNY) since 1965 and presently is the Chairman of the Art Department. He is also on the Doctoral Faculty of the CUNY Graduate Center. He has published extensively on early Italian art.

Brian Martin works in the Department of Science and Technology Studies, University of Wollongong, Australia. He is the author of 150 research papers in the sciences and social sciences, in areas including stratospheric modeling, astrophysics, numerical methods, electricity grids with wind power, environmental issues, nonviolent alternatives to military defense, information technology, the politics of higher education and strategies for social movements, plus the books *The Bias of Science* (1979), *Uprooting War* (1984), *Scientific Knowledge in Controversy* (1991), and *Social Defence, Social Change* (1993). He has specialized in studying the exercise of power in the scientific community in a range of areas, including nuclear power, fluoridation, pesticides, repetition strain injury (with Gabriele Bammer), nuclear winter, and the origin of AIDS. He has been studying and opposing the suppression of intellectual dissent since the late 1970s.

Gordon Moran, an independent scholar residing in Florence, Italy, received a B.A. from Yale University in 1960. He is the author or co-author of more than fifty publications, primarily involving art history and Sienese painting, but also dealing with topics such as scholarly communication, academic librarianship, and academic ethics.

Dhirendra Sharma, former chair of the Centre for Studies in Science Policy at Jawaharlal Nehru University in New Delhi, received his Ph.D. in Philosophy from the University of London. He has taught at Columbia University and Michigan State University. He has been a leading critic of the Indian government's nuclear power and nuclear weapons programs. Author and editor of numerous articles and books, his major publications include *Freedom Struggle of Vietnam* (1973), *India's Nuclear Estate* (1983), *The Indian Atom* (1986), *American Empire Building* (1981), *Asian Science & Industrial Policy* (1992), and *India's Commitment to Kashmir* (1994). He has been editor of the journal *Philosophy and Social Action* since founding it in 1975, has been convener of the National Committee for a Sane Nuclear Policy (India), and in 1991 founded the Centre for Asian Science & Industrial Policy Research.

Index